brown dwarf

book two - duden chronicles

Lynn R.Miller

*"Speaking of doorstops, my grandparents' little home be-
hind the store in Dover was so flexible and out of plumb
that every door in the house had to have a doorstop.
Some rooms, particularly the kitchen, like a wooden
trampoline would groan under the big men plodding
through. And it was a curious collection. A big green
plaster toad or frog, several boat shaped old stove irons,
a painted yard gnome, a big wooden Dutch shoe full of
stuff, a cast iron angel, a fancy glazed flowerpot."*

- from Clownery by Paul Hunter

Brown Dwarf
by Lynn R. Miller

Davila Books embraces thin fiction and poetic presumptions. We value creative insignificance, beauty and regularity. We haven't been at publishing long but we've think we've found a way to be at it close to forever. We like being out here on the fringe, a little hard to find. We like it because the air on the fringe allows us to breath and think at the same time. If you've found us, we challenge you to extend the spiritual life of a tree by buying books from us and reading them more than once.

Other Books by Lynn R. Miller
Fiction: *Duden Chronicles Book One - The Glass Horse,*
Talking Man,
Poetry and Prose: *Thought Small, Elastic Signature*
Essays: *Why Farm, Farmer Pirates and Dancing Cows, Old Man Farming*
Non-fiction: *Work Horse Handbook, Training Workhorses,*
Horsedrawn Plows, Horsedrawn Tillage Tools,
Haying with Horses, Horsedrawn Mower Book,
Starting Your Farm, Art of Working Horses.

Author acknowledgments: I thank Shannon Berteau for years of editing assistance, encouragement and proofreading, also to Chris Nash for proofreading and Paul Hunter for always believing in the writing, And eternal gratitude for the patience of my brilliant and magnificent wife, Kristi Gilman-Miller. LRM

If this novel were to be cataloged using the Mayonnaise System of the Richard Brautigan Library of Unpublished Manuscripts it would be noted as: ADV - ALL - FAM - FUT - HUM - LOV - MEA - NAT - POE - SOC - SPI - STR - WAR

On the front cover: A 1962 oil painting by the author.
On the back cover: An LRM photo of a homemade flax harvester.

To the memories of my teachers, Ben Maddox, Tom Guinapp, Don Weygandt, and Ray Drongesen. To Ben for instilling the sacred love of story, to Tom for the keys to seeing what wasn't obvious, to Don for the music of what is ultimately seen, and to Ray for the ease and truth of useful communication.

"I don't need to confide in anybody." - the author

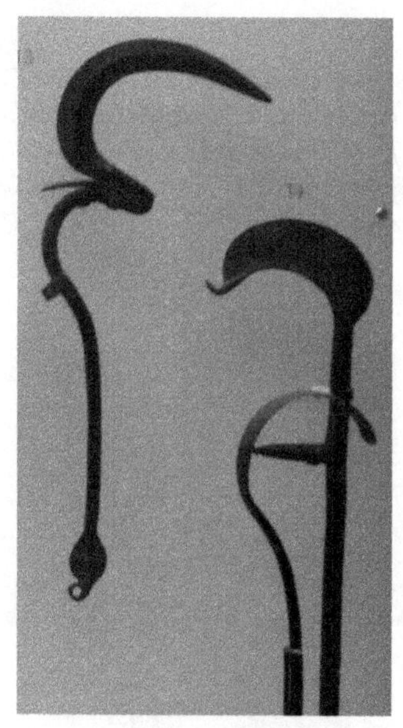

The images sprinkled throughout this novel are offered as
corrective clues less the reader find herself certain without
cause. Or they are random and have no substantive reason to
exist here. Or they are pieces of shed skin, evidence of depth
to the narrative and best secured by a driven ring-shank nail
piercing the pile of hides and anchoring same to that central
location just below the screw-driver rack on his shop bench.

BROWN DWARF
content

the Brown Dwarf's Dream

Some waking moments, horizontal with first eyes, useful and outrageous thoughts would flood his brain and his mind. At this age they could be expected to leave him early to his day. He no longer worried at their departure, he knew they would find him again if they needed.

And then there were other mornings, those when he sought the truth of the dream he woke from. Grasping memories, he would write down what he remembered and then, pencil still in hand, read it aloud correcting for sound, for sway, for expanse, for interiors, for the translatable details and for the sweep of it all. He needed to hear the before and after of the dream. On these mercurial occasions he did not trust the dream thoughts to find him later. Once done, he would throw away what he wrote and recount it out loud to himself, rehearsing as a practised liar might. This is what he told himself after last night's dream...

In an Imagined Future

He cut slices from the bruised, matte-skinned, rough gold of the Bosc pear and felt the glossy sweet of it stick to his old beard hairs. He was alone with this pear. His friends, the flies and ants and small sweet stealers, were gone. The tears he shed were colorless and evaporated before they could dry. All the signs, especially the absence of bugs, told him these would be the last pears for a thousand years. Scratching fountain pen and walnut ink on handmade paper, he was compelled to record these things for the deepest part of him required it. These were maps he drew not for his own undeserving kind but someday perhaps for a following kind.

The undeserving kind were beastly indolents in slow, deliberate pursuit of the doctrine of equitable outcome. They held oiled summaries of contracts granting them easy lives with thin, washable, globular shadows.

The following kind were those who held the handoff as sacred, who studied and committed to jalapeño memory the more powerful accomplishments of those before, who wasted no time with reinvention. They were certain that everything began when they stood courageously on the shoulders of those old ones they had learned to trust.

He lived in a vibrant, flimsy cabin on the edge of a vacant town many dozens of miles from any city. Didn't matter he knew they were all dead or dying, it had been coming for a long time; it had happened very fast. On one of his last visits to a bankrupt art supply store he had stocked up on paints and also purchased, for $1, a thirty pound box of plastic push pins - in mixed colors. His interior cabin walls, viewed sideways, glistened with the shiny pin heads which anchored the corners of hundreds of pictures some once printed from the internet. Later pictures were taken from book and magazine remnants he found at the landfill. First they were the visages of people who moved him towards hope: Pope Rudimento, Nadine Schwartz, Frederick Douglas, Bill Murray, Harry Nash, Alberto Giacometti, Charles Burchfield. And then he collected pictures of life forms threatened with extinction. Dead center of that extinction pinwheel was the heart-wrenching photo of one of the two last female White Rhinos adjacent to a starving Polar Bear. There were pictures of pristine wilderness threatened by human development. Here and there, a few pictures of conceptual art installations upon which he had rubbed actual chicken shit. There were pictures of lively small farm community landscapes adjacent to sterile expanses of industrial farming. There were pictures of wild cloud formations set against dull smog. And interspersed throughout were his own paintings.

He was alone with his animals, new rituals, evolving survival habits and his wrestling match with fear.

At one point his devoted Australian Shepherd, Geena, had disappeared. Gone for a week and he feared dead. Then one day she showed up and he was silly with gratitude. You might have thought him the dog for the way he wagged, he was that glad to have her back with him.

Then three of his small sheep flock died without apparent cause. His was a deliberate accident as stock farm. He raised livestock on small scale by choice and with glee. He found himself counting each animal every day as though that would protect them. They were all the company he had and some of them were his food.

Now these months later Geena, his stock dog, has had a litter of three pups and he dances a jig around her bed, while she day dreams of that wire-haired Terrier she had run with.

He has no electricity, none available out here. He has no motor fuel. No phone, no computer, nor television. He does have a solar powered wind-up radio and CD player. He never uses the radio but the CD player he rations for himself each Wednesday evening and Sunday morning. Camille Saint-Saëns, Mary Black, Jerry Jeff Walker, Chopin, Led Zeppelin, Paco de Rivera, Howlin' Wolf, Sam Cooke, Astor Piazzola, Bill Evans and Erik Satie CDs. His cabin is hidden in the woods outside of Seneca, so if anyone did happen upon the ghost town they would be hard-pressed to know he was there. That suited him. What few people he had come across these last years were desperate. And he had/has nothing for them. He has cleverly trained Geena to keep quiet and hide under the bed whenever anyone is near. He knew people were eating dogs now. He was rehearsing in his head how he would train the new pups to hide from humans.

Behind his cabin was an abandoned mine. It was large and long enough that he arranged his chicken run and corral pens so that, if threatened, he could quickly and easily herd, with Geena's help, all of the animals into the mine tunnel. All but the bees. The several oil-lamp lit, dry, cool tunnel reaches proved ideal for storing food and hiding valuables such as books.

Noah knew that the world hadn't come to an end. It has come to a temporary standoff. No way to know how long he has left. Maybe a year maybe thirty. But it was clear that the new deep silence is a biological death rattle. After the global depression and 'cleansing' wars, plus the destructive results of removing all restrictions from industry, new varieties of virulent flu had been aggravated by renegade cellular reactions caused by bio-engineering mishaps. The world's human population has rapidly been reduced by 60 percent. And, unknown to Noah, it continues to expire. Human fertility, depleted by personal choice, narcotics and virtual-reality bordellos, severely reduced the likelihood that humans would rebound. Tragic, but perhaps not for the planet.

Sylvan had walked from her stranded vehicle. She had walked for three days without food. Her eye glasses had broken when she fell. Feeling around on the ground, she found most of the pieces and wrapped them in a torn corner of her scarf. This she stored with her identification, map and the two remaining cigarettes in the big, mostly empty, purse. She figured that the town named Seneca had to be somewhere near.

She caught a light whiff of smoke and thought she heard a tiny yelp, a puppy's call. It triggered her hunger. She choked back the memory of lamb, fig jam, beans made with pork and molasses, fresh cherries, and port wine. The only way to hold the crazies off was to think deliberately and in complete sentences. Repeat those sentences, recraft them, pare them down to essentials. Sing those sentences out loud until they walk off. Tweak and re-tweak until the juice was gone. Make those sentences dissolve into the thinnest of perfection before morphing to dull and uninteresting. Her mind survived with this see saw, from the torture of elaborate desires to the numb shadowed far side of longing. The residue of her expensive education was the delusion that so long as her mind did not bring her down, her body had a chance of surviving.

〰〰〰〰〰〰〰〰〰〰〰〰

If the reading of writing is presumed to get one from here to there, the living landscape be torched of any meander or oleander, and any uneaten bits of the armoured toad's long dark wait should prove to never have existed, it would at-tach us after to the complete mossed evidence and grant what we call it to rise.

What does the reader deserve to know? True or false? Worthy or not? It be a writing - why must it be labeled as novel or no? It is novel. It may not be a novel. Close in, tight, that is where worthiness is protected and granted its place in the math of life. Out wide and scattered it can be nothing but the stuff of black holes.

Mumbles before clarity on one side. A need to understand on the other side. But mumbles before mumbles with automated code shifts by dead souls? Therein we have the big controlling rotting middle.

"The eye of the outward sense is as the palm of the hand, the whole of the object is not grasped in the palm."

- Rumi

Way back in time, to the present,
    where mistakes fester in their own juices
        until they become explanations for later
            when likely it will be too too later.

More of an actual Preface

complicated

chasmophile: a lover and seeker out of nooks and crannies.

the considerate lover's lasting reward

The dog shook his head. The invisible voice spoke a word, a dream-state word, spoke it right into the pear-shaped skull of Resumé, the old sheepherding crossbreed.

The Dog shook his head again. The word came again.

"Dog" was the word.

If the dog was capable of human speech his response might have been to say "What? What?" As in 'leave me alone.' As in 'I don't want to know what.' As in 'get out of my head.' But this was a dog too intelligent and cross-life pollinated to be bothered by such waste of essential energy. Instead, he shook his head, at least until some measure of understanding came to him. Or, if understanding got hung up and he tired of shaking his head, he sighed and waited stone-still for the next invasive picture-word.

Titus was trying to figure out this business of being dead but alive to, or at least aware of, the living world. Unless, was it the 'business of being dead'? Was this a demanding, post mortem dream-state? And to say he was trying to figure it out is not completely accurate. 'Figuring' didn't describe the urgency he felt with preparing himself to be specifically useful. Black and white, slowed yet still in-sync with the living world. Black and white, weaving, molten, through the many hued world, until a vibration set up just ahead of a definitive moment. No indication of what was expected, what was allowed; his hand out on the shoulder of a living one - felt the cold heat - absorbed the other's thoughts. Random was dulled. Touching one, he knew, should have brought more of a tingling. And touches were possible through a projected thought as well. It was resident with him, he needed to learn how to drive this stateless state of being.

So he chose the dog. 'Get to the dog' he told himself. 'Get inside of him. Pull his strings.' Zero expectation of anything other than some sort of trial run, it was to be a 'learn as you go' exercise.

Last night Enno had added the remains of his bean dinner to Resumé's dry kibble. This morning, at the same time as the word-picture 'dog' kept bouncing around in his skull, this canine felt a ticklish rush through the last inches of his bowel and he instinctively lifted his tail to make way for the passage of gas. "Ah..." He thought. Not really a thought. More an elongated, exhaled punctuation. The true thought that came to him was in the form of a wondering. Here was his thought:

'I wonder if the gas release has anything to do with the word that keeps popping up in my head? I wonder if, next time the gas comes, I can keep my tail down and stop it? I wonder if, at the same time, I can block the word from my head?'

Of course all of this has been translated from the haired-over Shepherd cranial wondering, a language without words. And you would expect not to understand. You might reasonably wonder what happened next. Actually it's important. It has significance to the long narrative, the blended margarita of romance and intrigue which follows. Here's what happened:

Resumé struggled, squinted, grunted without sound, hunkered down and clamped his short tail between his butt cheeks. And all of a sudden his ears popped as if he were in a rapidly descending plane. Not just popped, they actually squeaked. He got a whiff of a faint odor of rectal reflux wafting sideways from his cheekbones. His eyes went big as saucers. Right at that moment of rude surprise, against the sheer force of his will, his tail flipped up and a weak little trumpet of gas passed out his rear end. Then, as if to add salty insult to horseradish injury, the word 'dog' flashed before his mind in molten neon splash. All of this is evidence that we have been getting the translations right from the beginning. (Perhaps a little too bright, but right nonetheless.) Too many coincidences, you see, for us not to get it right.

(Besides, as the narrator I get to see everything, and, surprise surprise, arbitrarily tell you what you see. While it's true that I am not completely in charge of the vast landscape of small insults we might call the motives and persuasions of these characters and settings, I do get to tell the story. I get to make it true or false, thick or thin. And if I am lucky I get to tell it over and over again with florid variation and emotions respected. Parentheses in this book sometimes contain stuff you can skip over if you're keen to get done quickly. I promise they will not contain materials you need in order to figure out what's happening next; who gets the girl, how the bad guy is thwarted, or if humanity is saved. That all comes outside of the parentheses. The stuff inside of the brackets is proof positive that the narrator runs off at the brain.)

But as for whether or not this truly sets up the story I'd have to say not quite, for what we have here is to literature what the silent film is to the modern movie, what the old general farm is to factory agribusiness, what Mark Twain is to Malcom Gladwell, what the poem is to the text message, what humanity is to corporate, what a steaming cassoulet is to tater tots, what Robert E. Lee was to General Hooker, what youth is to earned crankiness, what unearned relevance is to obviated and excused passion, and finally what the individual *whine*-soaked and love-teased brain is compared to the lifeless google search.

"Tie chains to the back of the carpet, Joe, this way it will never fly."

Lloyd Jerald Shoulders, sculptor, rancher, hermit, sultan of the borderless lands, lifted his fountain pen and read what he wrote. Today's entry in his journal:

> *"those resonances which will not be owned or*
> *familiarized but rather which must own us."*

We, you and I, don't know whether to laugh at this man or fear him. He does, afterall, smack of someone dedicated to the seriousness of a felt mortality. Is that any way to live?

A little later, L.J. Shoulders was visiting with Sloop. They were talking cattle and hay. They were making circles in the dust with the toes of their boots until they realized it and wiped out the marks. Then they hunkered down and drew pictures in the dirt with fingers of choice. Up again in unison, standing with thumbs hooked in pockets, and finally arms crossed. Nothing they had said so far is worth repeating, it was the smallest of small talk, all clipped and abbreviated. Shoulders never could bring himself to ask a thing direct, seemed to violate the cowboy in him. Sloop didn't have that problem. He had the opposite problem. He said what was on his mind in a surgical manner, to cut in and draw out what he was looking for.

"What do you want Lloyd? Something's eating at you."

Lloyd made motions with his mouth as if he was chewing something, but he had nothing in there except what God and the dentist had left him with. Instinctively, he drew a bead on Sloop, looked him right in the eye, didn't flinch as he said,

"I had a good friend once. Never thought it would happen but we got stiff-necked and fists in pockets walked off from one another. Don't remember why. But today I avoid him like the plague. He thinks I'm a crazy old bastard and I happen to know he's a punk who'll never grow up...   I hate to see what's happening between you and the Duden boy."

"Leave it alone Lloyd."

Lloyd nods an invisible nod.

That's what he did. And glad to do it. Figured he'd made his point in drive-by fashion. Because Lloyd never wants to step outside his patterned and patterning life. Knowing what will happen next gives this hermit/artist/rancher his important security blanket. When there are spaces between spaces, where the brain skips a tone in its forward hum and Lloyd is forced to realize for that painful moment that **he** has prescribed his aloneness, he reaches for his notebook and reviews what he has set up to do next. Predictably the pain slides off like buttered eggs from a hot greased cast iron skillet and his chest hairs flutter. And all threatened depression crystalizes and falls to the floor. Now, there is a job to do. Now.

Back at his ranch house, Lloyd unloads the supplies he has purchased and then goes to the metal shop building that doubles as his sculpture studio. He removes a dirty bit of canvas tarp that covers a lump of something on a ramshackle pedestal. From around his neck and inside his shirt he pulls out what looks like a small magnifying glass on a chain. It's actually a reducing glass which he uses to gain some distance on the sculpture in progress, this one on the pedestal. It is a four foot tall metal and stone object combining free-form forge-shaped and hammered rods, akin to spider legs, along with a part of a discarded chromed bathroom faucet frame which has been threaded through a flat stone with a natural water-bored central hole plus three springs of different sizes, one enameled red. Looking at it from a distance he thinks he knows what else it needs and where he might find it.

Rattling the gate chain brings his saddle mare Juniper at a trot. He opens the gate to let her out loose - no halter no lead - and then shuts the gate so none of the others get free. He walks alone for the hundred and fifty feet to the stable shed. Juniper bucks, farts and scurries over to snatch a mouth full of yard grass before racing past Lloyd to the stable shed where she knows she will likely find some grain and hay waiting.

In the stall Lloyd puts a riding bridle with grazing bit on the mare and curries her before saddling. Behind the saddle he hangs a saddle-bag type affair made of four old cowboy boot-tops stitched together to hang, two to a side, like leather canisters. In one of these he puts a hacksaw, an expensive pair of German nippers, two vise grips, a ball peen hammer and two pigging strings. On the other side he puts a camera and a thermos. Last thing he does before mounting up is to tie a 65 foot lariat below the horn.

Leading Juniper from the stall, he smiles as he recognizes, as usual, that she is cranky. "Why can't he let me finish my hay?" she seems to ask. He gets in the saddle, patiently prepared for the little ritual of her crow-hopping and jiggin' before they set off towards the upper pasture and old ranch dump.

It takes him a half hour to find the cow herd, laying down in a draw, and to determine that all 85 horned Hereford pair seem to be quite fine and content. He notes that three of the steers seem to be considerably bigger than the others and writes down their ear tag numbers.

Back-tracking a bit, he goes to the hundred-year-old ranch dump and an abandoned wire-tie hay baler. From this he cuts a rusty three foot long, red, steel brace and takes a piece of chain off the tailboard. With these pieces secured in the pouches, he and Juniper head back to the shop/studio to work on the sculpture.

Somewhere else, seemingly unrelated, a whole 'nuther story tells itself; Loraine did it for science and she did it for the money. All those test rations they fed her during her pregnancy. Bland and harmless she thought. She had given passing thought to an abortion until the nurse had suggested that she could make upwards of twenty thousand by allowing her pregnancy to be part of an experiment. And afterwards, if she still didn't want the baby they would help her place it with a good family. So they fed her differing amounts of bovine growth hormone and various genetically modified organisms.

Then they told her she was going to have a boy.

She allowed herself to think about that for a bit. The baseballs, bicycles and frogs.

Then they told her they were mistaken, it was going to be a girl. So she thought about Oprah and George Clooney and mixed drinks.

Then they said they were wrong again.

"Damn, so, it's a boy?"

"We didn't say that."

"A girl?"

"Both."

"Twins!"

"We didn't say that."

"So, which is it?"

"So far the tests indicate both."

"Does that make any sense?"

"We've never had anything like this happen before."

At the time Loraine Frill was dating a Lebanese comedian who was reassured to learn that he was not the father of the test case. It was he who suggested she name the child Penobscot, that way as it developed it could go with either Penny or Scot as a handle.

And so it was born. As Penny grew larger and more complex he/she grew ambivalent about her/his dualized gender. This might have been tragic except that Penny was happy to be an 'it.' Happy to be gender neutral. Better by far than all the screwed up men and women 'it' came in contact with. Something about the moderated/blended hormones freed Penny to absorb knowledge like a chamoise. It also freed her/him to welcome psychic energy into 'its' very soul. Penny couldn't read minds, or see the future. Penny could, however, hear ghosts and see their shadows. (Some thought she could talk with ghosts.) With this combination of talents he/she had always been deft at deflecting ridicule. No one who knew Penny dared cross her/him.

Penny Frill was adopted by a sixty-five year old couple who owned and ran a car wash. From early in the child's development Joe and Meg Putaski ignored every effort of the scientific community to find and monitor Penny. They also avoided the many attempts by investigative reporters who had heard rumors that bioengineering had created a baby Frankensteinesque monster. The rumors and inquiries died down and out.

When Penny turned 13 he/she had enough credits to get an advanced college degree. That was good because Joe and Meg were old and delirious and could no longer take care of themselves let alone Penny. So the mutant marvel got a job at the Mascara library and took care of his/her adopted parents. Penny loved the library because it was a favorite hangout for ghosts in transit. It was like a rest stop on some celestial highway.

Penny carried a big handbag in which resided his/her mute Chihuahua/ Lhasa Apso cross doggie.

(I was doing research for this long story, walking the rainy streets of Astoria, Oregon, when passing a nasty, 'we are hipper than you are' bookstore, out front two plastic chairs both occupied, I heard the seated small people of male persuasion, I think, speak to one another.

"I gotta try some Opium, Jeck. I do."

" I don't need a mountain man."

"Did you hear me? I gotta try some Opium, Jeck"

"Don't make me hurt you."

Looking up at me as I was looking down at him, he nodded his small head like a pointer,

"This big ugly man walking by and pretending not to listen to us, Yes, you. He wouldn't even stop me. So honey don't, or I will hurt you in ways you will not be able to enjoy."

I thought, now this is something that might lace up my story but how do I use it? Well, I answered myself, you use it to prove that you just can't make up this stuff ... it comes at you, pure. Blind we are to treasures floating 'round our ankles. Perhaps it is such asides which are the real story, part of a mixed pile of crumbs on the edge of this table that we like to think of as a novel)

The earth's skin stretches over a slow moving ocean of competing rock slabs injected with hot and cold water, gases, and petroleum ooze. That skin rises and lowers, undulates, cups, breaks over ridges, fluffs itself 'neath forests, pats itself down under seas. In many places and ways, seams can be found where the blanket skins of the earth are stitched together. Hidden from any realtor's understanding, the central reaches of a state in the region erroneously and temporarily known as the Pacific Northwest, is a vast boundary-less territory comfortably regarded for over two centuries as Oregon. The moldy western edges of this territory absorb the Pacific's occasional deep fogs, but it is the Cascade mountain range which offers the best defense against the tubercular invasions. The weather changes on the eastern side of these mountains. It's drier. Much drier. So dry that most may be forgiven for missing the seams. Here in central Oregon the subterranean forces of the living earth are very near the surface. And the Cascade mountains, these young and fractious volcanoes, are the living pimples on the territory's skin. At any point they might burst forth. But for our story, more important is the fact that the root structure of these pustules give access to secrets that just might permanently change the human landscape.

Roman concrete - lime plus volcanic rock mixed with mediterranean seawater causing a chemical reaction in which the lime incorporates molecules into its structure and reacts with ash to cement together.

The crystalline structure of tobermorite.

Enno kept to what he felt was his damaged core. He did not have to answer to anyone. There it was. He was mortally embarrassed and conflicted in his suspicion that no one was more intelligent than he. Regularly he rejected such thoughts. He understood there were those who knew more things, were far more capable. And then there were those, above and beyond, who were far wiser. But when it came to native intelligence, he felt himself to be the equal, even superior, of any and it pained him. Which is why he was also certain he was as dull, as unintelligent, as most. It was this contradiction of certainties, deep in his core, that held him in sway and kept him, in all ends and in every time, as humble as meek might define. Never to parade, he however would and could murder with his thoughts, and this knowledge would have him murder himself. None but the coming dwarf would ever know this, and understand that the balancing, hovering evil which gives goodness its depth and value, stained Enno's power. Together, both of him, all of him, they watched as Enno hung back away from his outward self and bid his time.

drunken heads
piked
shadows confused
the sculpture of it
never forgiven.

shaved go the bottoms

Part One

Alforhas

Babylon's Ishtar Gate

Sir Walter Scott wrote, "the magic power of imposing on the eyesight of spectators, so that the appearance of an object shall be totally different from the reality."

"In painting as in music and literature what is called abstract so often seems to me the figurative of a more delicate and difficult reality, less visible to the naked eye."
- Clarice Lispector

Chapter One

the sand lily's pause

"I think that a lot of creative people never grow up. I am certain that a real man wouldn't paint any pictures! Or wonder about the universe, or believe in dreams. Or think that trees sometimes look at him."
- Willem de Kooning

Brown Dwarves? *These are the "failed stars;" objects with too little mass to fully ignite nuclear fusion in their cores. Instead of blazing with red, yellow or the white light of our own stars, they're heated by the gravitational collapse of material.*

As he watched the old stock dog, 'Resumé,' lift his leg on a rusty remnant of single-walled, steel stove pipe, he wondered at the age of these homestead remains; seventy or a hundred years? Certainly no more than that. This, even as wild and western as it felt, this was new country, much of it homesteaded as recently as 1920. And the rock wall was a haphazard piling of field stones, which zagged past the old hand-dug well-hole, on its way to the now seemingly ancient cottonwood trees, lumpy and shaggy in resignation. But they weren't ancient, those cottonwoods. They were old, yes, as old as the homestead, old as their straight-stalked poplar cousins, but not ancient. The true ancients here weren't even the scattered old Pondersoa pines, the true ancients were the Western junipers. Enno had no way of knowing that the big, twisting, gnarled scrag of a tree he looked up at, with clumps of tangled needles in two bunches near its broken top, had first sprouted from the ground in the year 1066 A.D. And he had no way of knowing that this self-same tree, ninety years past, had been a post around which a farmer had strung steel cable to anchor his ox-powered stump-pulling operation, in this way to clear the land around that homestead. And he had no way of knowing that the broken stub of a limb, twelve feet up on its trunk, told the story of the failed attempt to hang an overweight horse thief in 1904. And all records were lost to corroborate the native's telling that Captain Fremont had won a urinating contest with a tribal elder, each aiming to mark, with their water, a highest point on the old trunk - this following the intake of three-fifths of a hair-tonic-thinned alcohol.

Many of the locals, and especially the most recent crop of government bureaucrats, all publicly detest the junipers, seeing them as intrusive weeds and water thieves. It was true, that this wide, high desert landscape, before the Europeans came barging in to lay claim, burnt over frequently - sometimes when natives would set cleansing fires and often naturally by lightning strike. The aromatic juniper provided oily tinder, fueling hot fire centers as the sage, bitter-brush and bunch grasses raced to form lacy ghost-writing charcoal. The youngest junipers seldom survived. These fires prevented this member of the cedar family from 'taking over.' Fire prevention and 'civilized' man's designs had altered all of that, making of the juniper an easy and defenseless foil in the shift of blame for unwelcome landscape changes. But Enno Duden didn't know any of it, and doubtless wouldn't have cared much if it were told to him. He admired the young and old juniper as he did all trees and felt the need to give long and careful consideration before cutting any one of them down for whatever the purpose. He wasn't soft, he wasn't a tree hugger. It was more like he felt it was his duty to give pause.

And it was a pause which caused him to notice the washout of the original small pond. Odd, he hadn't noticed before, the volume of water rushing through that eroded ditch. It was like a small stream. But there had been no surface water on this property, not when he had first walked it. And recently there hadn't been any late snows, or heavy rain to account for a runoff. But then he hadn't walked it all. He couldn't, it was too vast, with canyons and swales and rises concealing the short view. He had, however, walked this old homestead site and he was certain there had not been running surface water here a week before.

The hurried water ate away at the sandy anchor-soil crowding the roots of another twisted old juniper. It was a short tree with a massive ridged trunk, five foot through at the base and tapering rapidly to a narrow top, just twenty feet from the ground. Midway up the trunk, a tousled mass of fine branches exploded in a confused mess. And framing those thin curving sticks were fatter limbs covered in a chartreuse-colored lichen. A short distance away, a cluster of aloof young Ponderosa pines shot straight up fifty feet and higher. The short, shaggy old juniper was a friendly witch of a tree compared to the arrogant, tall, small-brained pines. The juniper talked to Enno, telling incomplete stories meant to end in laughter but never making the full turn. The pines held their noses up high and whispered amongst themselves. The junipers were also very different from the Douglas fir and alders Enno had known on the wet side of the Cascade Mountains. The fir were lofty and the alders were sour lemony trees of short life and shorter tempers. The junipers, on the other hand, were comedians and always accessible, always ready to lip-sync to any imagined, wandering wood nymph's slide whistle melody. Enno Duden didn't think about these things. Without thinking he felt them. They made him happy. They added a flavor, a justification to his pauses.

He held on tight, it was happening so fast now. The external world stood still while he whirled at his core, an emotion-driven gyroscope. His head felt as though it were floating or bobbing separately and just above. For all of his long young life, with a powerful certainty, he had held onto his one dream, his one goal, through every distraction, every disillusion, every misstep. Someday he would have a farm of his own, and now that moment was within his grasp. Now he was being given a chance, an affordable chance. A farm of his own? Nettie and Sam had made the arrangements for him to purchase this land with graduated payments, no money down. For the first three years it would be only $100 per month.

(He would never know that Nettie's preference had been to give him the land outright. Sam was the one who insisted that Enno *pay for it or it wouldn't mean as much to him.*)

He would continue to spend each day exploring the property until the attorney asked him in to sign the papers. There was that escrow business to get through. Thank heavens he had Sam's cabin to use in the mean time. Not much money. No job, no way to pay rent. But so what? For now the future was all ripe good luck and he young enough to presume. He had forgotten, for the moment, how it was not long ago that his friend Sloop had driven him off in anger. He had forgotten about that tragedy which had taken the old man from them all. It didn't seem to help that it had been Titus Ibid's choice; he had set off the explosives to save the rest of them and Jefferson's farm.

No, today, this morning, Enno didn't want to think about any of that, he just held on to this realization of his own great good fortune and itched to get to the enormous job ahead. For him celebration and planning were one and the same.

And now it was time to return to town. Back in his pickup, dog by his side, he turned on the AM radio. The news broadcast reported that plain-clothes policemen with a warrant had knocked on an Atlanta, Georgia, residence door, announced themselves and, with guns in hand, broke the door down. They were met by gunfire from the frightened 92-year-old female resident who wounded two of the policemen. They returned fire and killed the woman. Family members said the woman was frightened by her neighborhood and kept a pistol for self defense. Police said the raid was drug related and offered no comment other than the 92-year-old female was accidentally killed. Duden closed his eyes in prayer. His happiness slid from him like a disgusted, paid lover. Now he was exhausted by complicity, for every human tragedy made his young self ache. He would stop it all if he could. But he could not because everyone was wrong, everyone to blame, even he, especially he.

He had gone to the woods, to the northern desert, to the far unpopulated reaches, looking for a piece of land to farm. He needed to be away from any population center. Though he was young, the paradox did not escape him. For the many, leaving behind the security of human confluence for the uncertainty of a threatening wilderness was ridiculous. For him, human confluence was the threatening wilderness. He did not think about these things, he felt them. His was the purest form of emotional intelligence. With little formal education and wholly inadequate social skills, he found abiding comfort in the single-mindedness of his purpose, in the solitude of all that. He was no idiot. He was his own instrument, his own tool, and he knew, emotionally, he was to be a farmer. In these ways he was a beast of welcome burden, an animal, unencumbered by wasteful peripheries.

After the recent cataclysms: the deaths of Billie and Titus, the loss of his friendship with Sloop, the disappearance of the one young woman he had allowed himself an emotional investment in, and the mayhem of the volcanic eruptions and confusing cross-fires, Enno fell back in step with his dream. And now, without explanation he had been given a chance at this remote high desert acreage, once long ago a homestead farm, now desolate and completely disconnected.

The realtor had called it Snake Flats. Narrow rough dirt roads, impassable in the winter, wiggled and waggled without warrant through the juniper and pine thickets, crossing miles of Forest Service ground before reaching, passing through and leaving what was soon to be Enno Duden's land. He was afloat with the ecstasy of plans and planning. He knew, or thought he knew where his cabin would go, where he would build his first little barn, how it would be situated off the corner of the old abandoned hayfield. He thought about how he could incorporate part of the old stone wall into an enclosure for pigs, and further down where he would set up a night yard, inside that same wall, for his small band of sheep. He thought about where he might clear an area for a horse and cow pasture so that it would be easy to get them into the barn. He tried to imagine where the sunlight would come from during different times of the day and seasons. He wanted a greenhouse right away. It could be crude, just good enough to extend the short season and supply him with vegetables. Later he would build an ambitious one. So many thoughts, plans, dreams, adventures, all of them leading him to that self-sufficiency he craved with a religious fervor. But now, set the dreaming aside, it was time to go, he needed to get to town to take care of business.

This time he decided to go down a different dirt road to see if it would hook around on the secondary county road. Before the trail left his property, he noticed, straight across, the forty acres of pine tops and realized that these trees were taller than they appeared. They all began down in a small canyon he hadn't seen before. He instinctively inhaled and even from the distance felt the dry rosinous air creep up into his nostrils. These were fine big pines, what the local loggers called old *pun'kins*, an acknowledgment of their wide, round, orange trunks. Enno thought he would explore that stand on another visit and then he saw, down in the canyon, a glint of silver trailer wall. Resumé, his old stock dog, growled from the truck seat.

"Whaddaya see Rez?"

Enno stopped the truck and they got out, both man and dog peering down into the thick stand of trees and looking wide for some sign of life.

"Hello?" he hollered, as he hit one short blast on his truck horn. He heard a dog bark from below and his own canine answered, hesitantly.

Enno walked down towards what he had seen. Threading his way through the trees he made out the shape of a rusty van hooked to an old

Airstream trailer. Resumé walked just back of young Duden. The boy knew that the dog would risk his life to protect him, but only if it was absolutely necessary. This dog was no fool, long ago he learned not to rush into something that smelled wrong. This was not necessarily a wrong smell, but it was a curious, confusing, incomplete smell.

Making out the shapes as he approached, Enno thought he recognized the two dogs which were tied to the Van bumper, then he saw the form of his old bedraggled, stuttering, homeless friend, Jimmy Three Trees, squatting by a small stone structure, maybe eight foot square. Jimmy had hold of a rope which ran up and through a big, double snatch-block suspended from an odd, heavily timbered A-frame which seemed to sprout out of the roof of the stone building. Where Jimmy squatted he seemed to be looking through and down the doorway to the building.

Sensitive to his fragile, jumpy, easily disoriented friend, Enno knew better than to frighten him, so he stopped, stood still, and watched from a short distance. Jimmy's two dogs barked and Resumé growled but the old man ignored those sounds and kept staring down, all the while holding the rope, doubled around his waist and taught. Resumé heard it first and lifted an ear on a cocked head, a whistle blast, then another. Jimmy stood up and pulled on the rope, hand over hand, backing while he did so. Finally, rope going slack, Enno heard a voice and saw a tall, thin man with a lit miner's cap. The cap was crowned with what looked like an ornate, miniature, metal, victorian brazziere. The man came out through the stone building's doorway all hunched over. He shook or shuddered himself. Then the man in the apparatum hat saw Enno and nodded to Jimmy who turned around. Enno Duden in his sights, Jimmy smiled wide enough to swallow an ocean sunset.

Introducing himself, Enno learned that Jimmy was working for Dr. Fil, the nickname preferred by Dr. Filipiano (the world-renowned Nobel Prize winner who had dropped out of sight years ago after a protracted argument with his bosses at the Haptic Institute for the study of Abnormal Geophysics. Dr. Fil's "Lily" theorem assigned the big-chair, small, old, body, disrespectful brat aspect to 'Brown Dwarf' inquiry.)

Enno learned just the name and the fact that this was a scientist who studied dark energy and gaseous bodies. He sighed. This was another diversion, and a harmless kook.

Jimmy repeated the introductions, including Resumé, this time mentioning that Enno was the new owner of the land they were on. Then Jimmy took the piece of bubble gum he was chewing from his mouth and offered it to Enno's dog who gently but quickly took it and trotted off behind a tree.

Every once in a while, Duden was surprised by evidence that Jimmy, had steady, deep interiors to his fevered brain, interiors where he gathered and held information and observations.

Dr. Fil wore chest-crossed ammo belts from which hung tiny, lidded, mason jars full of what looked like floating, phosphorescent, lint particles. He also wore the smile of one whose private joke has sprung a whole new punch-line.

"This is quite a unique piece of property you have. Are you interested in selling it?"

Duden frowned and gently shook his head no.

"Ok, we do need to talk, there are things you should know about this place. Are you familiar with the 'Kessler Effect'? No? Well, it applies to outer space phenomena and I believe I have discovered how it may apply to subterranean forces and conditions. That recent earthquake we had has caused a whole lot of underground 'smash-aparts' as I call them. Below our feet the ground is changing dramatically and pressurized water is making its way up to any weakness in its surface. Hot water, cold water, and waters laced with valuable tailings. You may have, no please let me say it, you DO have a gold mine here."

The look on Enno's face was that blank, oily, stillness fishermen know as calm, deep water. His eyes went slowly from side to side, first to Jimmy then to Dr Fil. The dog read it and knew that emotionless Duden was worried. Resumé rose to his part in this moment and gently took Enno's loose hanging right hand into his mouth and carefully gripped it with 'I love you and everything is ok' pressure.

Loading his van Dr. Fil said, "We are done for today. Pulling out now. I trust we may get together soon and talk about permission to continue access to this shaft? For this I would gladly pay."

Thoughts of how the stories, which fill with details and slow to anticipations, follow; those are the ones with energy.

"Stay in the vehicle." There it was, nasty, absolute.

"Keep your hands where I can see them."

"What's going on here? I didn't see you behind me."

Lights flashing, the patrol car was stopped across behind his parked pickup truck and blocking the Mascara side street. Left hand on the butt of his holstered revolver, right arm extended, palm face forward, he said,

"Slowly and carefully pull your driver's license out and hand it to me. No quick moves."

The old man reached forward to his dash and got his wallet, opening it and sliding out the license and his insurance card.

"Just your driver's license, pay attention now," came the warning.

"What's the problem? What's going on." He started to slide out of the open truck door.

"Get back in there and stay put." The officer backed deliberately to his patrol car radio, never taking his eyes off the old man.

When the officer returned, his walk was relaxed. "Here's your license back. I am going to have to cite you today. It is the law and there is no excuse. It's going to be expensive but its the only way people like you are going to learn. Next time make sure your seat belt is fastened. There is a court date on the ticket if you wish to contest it, but I suggest you don't even think about it."

The officer strode back to his patrol car, shaking his head at the two bags of trash and wet loose hay in the back of the old man's dirty pickup.

The old man weighed whether or not this was one of those battles worth the trouble.

Today most people under the age of fifty are resigned without panic to the notion that humankind faces likely extinction. 'Hey, if the planet's gonna die, we go too, don't we? Pass me a plastic fork, will you?' It is a most remarkable and insidious side effect of human immersion in the cyber universe, this buffered, suicidal indifference.

Nostrils wide, muttering, we go backwards through the turnstiles of human evolution, passing the piles of dead batteries and collapsed promises, all the way back to the toxic muds.

History does not yet know its good fortune to have strong handfuls of simple-minded, purpose-driven, land-loving, ocean-adoring, tree-hugging, dog-following, sky-worshipping, bean-growing men and women who refuse to ride the gas-filled balloon of civilization's assured obsolescence.

> *the earth moved*
> *askewed pattern*
> *laser bullets*
> *so precise*
> *mathematician's nod*
> *though target lives on*
> *world quakes*

Chapter Two

the calendula's elbow

"I was beside myself with ineptitude, and I felt so ashamed
I just wanted to go home and behave myself."
- Karen Elizabeth Gordon

Back in the town cum city of Mascara, Enno went to the auto parts store for a new set of plugs for his truck. Maybe a condenser and points, if he could afford them. Barely in the door he heard the sideways greeting.

"Well, if it ain't the farmer boy."

It was Gunny, a red-faced, red-haired, blond-eyebrowed, lobeless malcontent. Gunny Sax was his real name, actually Gunnison Tenner Sax, and he ran the store. He used to own it, but got burnt out. That's why he decided it would make sense to sell the business, only thing was the people who offered him anywhere near what he thought the joint was worth required that he stay on and run it. So, practical man that he was, he first got himself an attorney - a good one (say expensive) from Bend, to deal with the commercial realtor and the buyer and the buyer's attorney. Then he got a good accountant (say creative) to help him with the tax questions. In the beginning Gunny felt pretty good having all those well-dressed profession-als sit down on his behalf. At home he told his depressed and depressing wife that everything was gonna be different because they were selling out and there'd be money to do things. And he explained how he had negoti-ated a slightly lowered price if the new owner would give him four weeks off before returning to management.

That set his wife off on a tear that never ended. By the time the yell-ing had stopped, a week later, she had an attorney of her own who filed for divorce on the grounds that her husband was an incurable and unbearable idiot. So the escrow of the store sale had a lien slapped on it for property settlement. By the time Gunny had signed the papers, his ex-wife-to-be, his attorney, the buyer's attorney, the divorce attorney, the realtor, the ac-countant, the escrow service, the title insurance company, the sandwich shop across the street, the office supply store down a block, and his ex-wife to-be's dog groomer got almost every penny of the money. So Gunny found himself with a four week holiday and two hundred dollars in cash. Practical man that he was, he bought a fishing license, borrowed a friend's boat and went to Lake Billy Chinook to troll for Kokanee, those lovely land-locked salmon.

Running the boat up on rocks at dusk of the second day, he found himself sinking slowly as the welcome sight of a tribal patrol boat rapidly approached. They couldn't save the sinking boat. But Gunny had his three buckets full of fish... 'ceptin' that the game cop was from the Warm Springs reservation and Gunny did not have the requisite tribal fishing license. Choice, and not so choice, words flew, and Gunny spent a week of his holi-day in the Warm Springs lockup.

Selling the shop was to be the great "fix-everything" for Gunny. Now, all this advanced bad luck had freed Gunny to be the full-blown obnoxious malcontent he was born to be. There was little or no reason for him to be polite, considerate, appealing or hesitant. He had job security because he knew the business and because no one else would work for so little pay. Or so he believed.

"So farmer boy, how's life on Snake Flats? You figure out yet how to make a fortune off of that rock pile? Let me guess, you gonna plant a very big rock garden? Or maybe flood the place and grow rock fish?"

Gunny had a silent laugh. No sound passed his lips, but his shoulders shook up and down like a rolling body hiccup and his face flushed an even brighter shade of red than usual. Right now he found himself terribly funny. Gunny didn't notice Resumé, in the cab of Enno's truck, staring through the shop window at Gunny. The dog alternated between a grimacing, low growl and an antic, gum chewing motion.

Enno tried ignoring Gunny. Eyes down, matching his breathing, he nodded without commitment. Couldn't be seen, but it was there. Sideways, Enno Duden made out the form of a second man sitting at the stool on the customer's side of the counter. He kept the man's form on the edge of his periphery. If Gunny Sax was going to be a jerk, Enno wanted to be as transparent and thin as possible, invisible even, and for him that meant pulling way inside, sucking his eyeballs back to behind his jaw hinges, avoiding, at all costs, any eye contact. He sought inconsequential. Not out of fear, it was his form of polite dismissal. He had pulled back from the counter to a display rack and fingered through some engine additives and picked up a bottle of Mystery Oil to read the label. That word, Mystery, took him right back to the conversation with Dr. Fil. Withdrawn by choice it was an easy matter to slip back an hour to what had just happened...

Dr. Fil. What a strange one that guy had been? 'Seem to be attracting a lot of strange people these days,' thought Enno. This one, looking like a cross between a young Laura Dern and an old John Carradine, apologized for trespassing and said he was just gathering dust particles from down inside the old well hole, putting them in the mason jars to take back to his laboratory. Something about the dark energy of inner space. Not being the actual owner of record yet, Enno Duden had avoided pushing any claim. Anyway, this guy seemed like a harmless kook, perfect counterpart to Jimmy. Duden probably wouldn't have argued about trespass because it wasn't his nature. And Dr. Fil didn't feel like he was trespassing, maybe partly because Jimmy Three Trees seemed so fond of him. For Enno, Jimmy was more than ok, he was a fellow insertion in a sick and cloying world. Jimmy didn't say much but when he did it was like a diuretic stream of random associations, he couldn't stop without clamping his hand over his mouth.

"Boy says we trespass but he don't say it cuz cuz cuz he's a k.. k.. kind one, the ones who pa.. pa.. pa.. pat at those clouds and bunch 'em up fluffy or never squash bugs cuz cuz cuz cuz they may be wa wa wa one of us back again or because he smells things like like like like some folks sees things, things in the shadows or or or or out where the girls spre spre spre spread skirts over updrafts and gi gi gi giggle diet giggles and ra ra ra rust boys pipes with empty pra pra pra promises. Pipes is big straws, really bi bi bi big straws, and we we we we should stick wa wa wa one in the top of the capital and sa sa sa suck out the silliness of our con con con congressmen right up through their nog nog nog noggins or out their ear ear ear ears because they only hear what they want us to ag ag ag agree to..."

Jimmy had just kept on muttering. Dr. Fil spoke, over the top of the mumbled, garbled, word stream. He spoke of his 'project' in guarded terms. He fell to genuine comfort when it became obvious that the boy, Duden, didn't have a clue what he was talking about. As they parted company Dr. Fil, over his shoulder added "Oh, and get ready for that water."

Back in the parts house, Enno's reverie was disturbed by Gunny coughing up another phlegm-coated glob of abuse. "So, seriously kid, you've got to realize that in order to farm a man's got to have farm land. That crusty piece of crap you landed ain't even gonna grow food for scorpions."

Gunny needed Enno Albert Duden to join him in the camp of congenital losers. Then, from the dark man on the stool came a whispered voice, husky and percussive mixed with a harmonic whistle over top.

"Farming's a fine art. Nothing short of that. And art is all about making something from nothing. Give the true farmer, the farmer at heart, a 'crusty piece of crap' as you say and he's way ahead of the game."

Gunny glared at the grizzled, little, dark-skinned man seated on the counter stool, then turned on his orthopedic heels and walked down the parts aisle. Enno first noticed the interrupting man's fine, polished, bulbous nose, it came out front of the head distinctive and announcing all the same. Beneath that there was a scattering of fine white beard stubble. Capping the older man's head, as if to prevent him from ever gaining any height was a curious, rolled, leather cap with a torn edge. It was the man's eyes which stopped Enno. Eyes which apologized in full for every intrusion ever made by any human upon another. And eyes which held a perfect balance between serious and hesitant. Enno found himself nodding and the little, old man nodded back.

Gunny came back and pulled a motorcycle part from a plastic bag, slamming it on the counter. "There it is. Last one we'll ever carry. You take it or leave it. No refunds. Want it?"

Benji Aboo pulled the part towards himself with one dark, leathery finger while keeping an eye on Gunny. He whispered "How much?"

"Twelve fifty."

Benji pulled an envelope from his coat pocket and extracted a ten and a five. Enno found himself watching and feeling helpless to look away. As Gunny made change he spit out sideways, "So whadda you want farmer boy?"

Enno started to tell him and then, invisibly, fumbled and stuttered as he watched Benji get down off the stool. The old man stood only four foot tall. Tall is the right word, because he didn't look short, he just was. As Enno watched the man leave the store he heard Gunny spit out "Creepy little black bastard. Who the hell does he think he is anyway? Now what was it you wanted kid?"

Enno turned slowly to face Gunny. He took a guarded breath and painfully remembered what his impulsive action had caused Helga. Rolling his jaw once, he spoke in the loudest, slowest, whisper. The words were drawn out, stretched as though coiling before a strike. It was as if a cobra were breathing into a microphone.

"I don't want anything from you. Nothing."

And Enno turned and walked off as Gunny cursed to himself in beer-latin. Outside of the auto parts shop, Enno found the little man standing by his truck and petting Resumé's head at the open window. Benji had to reach up to the gum chewing dog.

"You are Enno."

"Yes, but how did you…"

"Shirley told me about you."

Enno felt weak hearing her name. He also felt nervous. Apprehensive.

"She asked me to check on you, to make sure you are safe until she returns."

"She's coming back?" His own words sounded more like an announcement than a question. Enno was guarded and elated.

"I could use a lift to my bike, it's on the other side of town," Benji said without answering Enno's question.

The tall, young man behind the big black wheel of the old Chevy pickup; an interesting silhouette set against the erect stock dog and the little man, who in the wrong light looked like a whiskered, black child. In the right light he looked and felt Tibetan.

Enno learned his name and that he had come up on his motorcycle from the California Redwoods where he lived in the forest. He had come because Shirley Intoit had asked him to find Duden and look after him. The motorcycle's starter switch had broken and it was purely by chance that Benji had run into Enno at Gunny's shop.

"Well, nice to meet you but I don't know what she's talking about." He stroked his dog, "We are okay, we don't need to be protected."

At that very second a bullet passed through Enno's passenger-side window, just ahead of his nose, and ahead of Resumé's stock-dog ears, and above Benji's cap and out through the passenger window. Slamming on the brakes and veering to the right, Enno accidentally hit the horn, then the curb.

In the darkest night he would go with Phillips head screwdriver and half a dozen 'parking reserved' signs. At all three banks, he affixed the fake signs telling the whole world that these here parking spots were reserved for Bo Coin and that all others would be towed. He was painfully careful to make sure no one saw him do it. He put two up at different curb spots on the main drag. Then, as pièce de résistance, he put one up at the post office parking lot, right in front of the door.

My gosh but he hated the realtor Coin. He hated how Coin had cheated his widowed aunt of her property. So in response he began what would surely be a long campaign, by 'helping' Mr. Coin, and his arrogance, bite himself and itself in the butt, over and over again.

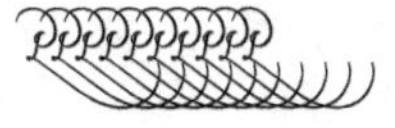

thorny elapsicans devour
multilingual pauses
subliminal sauces
and water core apples
while thankless children
do the coin rub
across slow narratives

The magnificent, storied, **Royal Portuguese Cabinet of Reading** (Portuguese: *Real Gabinete Português de Leitura*) is a library and lusophone cultural institution. It is located in Luís de Camões Street, number 30, in the center of the city of Rio de Janeiro, Brazil. It contains several volumes by the great-great-uncle of Benji Aboo, Fernando Pessoa.

Chapter Three

and the bunch grass brushes

*Law, in part, developed out of the social impulse to
minimize the collateral damage of the taking of revenge.*

*Ah, the stupid tyranny of symmetry, the circular insanities,
and the coin of the manic-addictive*

"I'm like a slipping glimpser."
- Willem de Kooning

Titus didn't know he was Titus. Not any longer, not in the
people way. Titus wasn't a person any longer. He was a presence, a floating invisibility. His spirit had no mind, it had
intent, it had an untagged memory, it had energy, it recognized but without
catalog. Titus knew he had things to resolve, things he needed to correct,
things he needed to influence, but he also knew he wasn't to influence
anything. Afterall, he was what some call a ghost, others a spirit, others still
more exotic names. And there were rules to his conduct, rules determined
more by gaseous molecular edict than by codebook. This he learned as he
found his effort thwarted in one direction and rewarded in another. But
then how could he have known or learned if he had no mind? We aren't
to understand these things except as the contradictions they are. And we
are to understand that the words which struggle to explain are tiny ridiculous things only granted poetry and meaning in the slingshot/lazy-susan/

mental-negligee of the human monkey in transition. Need it more direct? Titus wouldn't know what to do with a cheeseburger and he wouldn't care. Racism, sloth, greed, and the hiccups do not exist for Titus. Titus has no credit rating. But Titus could see what he wanted to see.

He could see Enno Albert Duden and he weeped to see him, because now he knew that the boy was fresh in those ways he, Titus Ibid, had once been. Fresh, not pure. And now it was so clear to his transparent wispy self that fate is actually just an excuse for the weak-willed. Now he knew that stoicism isn't strength, strength is having the lubrication of sentiment to follow the ticklish impulse.

He knew he couldn't stop the bullet from being fired. He knew that whatever he did had to be thrice removed and of only the slightest conse-quence. In the light breeze that is his transitional condition, insignificance must be worked at.

With every ounce of his being, Enno found himself pushing back deep against the truck seat. Looking sideways he saw Benji sitting calmly with his eyes shut. Resumé was shaking.

"Are you okay?" Enno asked them both.

Resumé whimpered, growled, woofed and stuck his nose in Duden's armpit. Benji opened his eyes and, looking straight ahead, said:

"You have another friend, one who pushed me aside. My guess, it's one you knew as Titus, Shirley spoke of him."

"But..."

"I know, he has passed on, but he is here with us now. And he saved you from that gunshot. Life is only sometimes tailored, sometimes it is ragged, loose-fitting and subject to multiple choice."

Titus, who has no ears, heard Benji's words and they made him mad as a child. Titus who has no body felt himself shake. And without explanation he knew that Benji's prescience threatened him in his work. This man was a seer, or as some might say a good guesser, and Titus had to be careful, or this man would have Titus taken off the job. Titus Ibid, the ghost, now had to go away and find some other appropriate challenge. Somehow he had to devise a way to be working through another entity, once or twice removed, before he returned to the boy. He felt himself tighten, claws extending, like a cat being licked by a puppy. Being a ghost was a bitch or it made a bitch of you.

"There may be police," Benji offered as they walked around the truck to see the damage.

"I hope not. The only damage appears to be the holes in each glass. It had to be some sort of freak accident." Enno didn't believe that, not for a second, but he wanted to. He wanted all of the intrigue with Sloop, Sig and Titus to be over. He caught Benji watching him.

"We better move on. Take me to my motorbike, please." He said this with some urgency.

It must be held that man's dependence upon the natural/biological world is both his blessing and his frailty. To separate man from the biological world by industrial circumvention and/or Frankensteinian manipulations is to ask of humanity that it live as a species immersed in the incubation of disease and clothed in the opaque, craquelatured nightshirt of the dispirited.

If Menand's "Mosaic Law" instructs both action and belief, would such an observation also apply to the construct of two dimensional art? Would not the mosaic of a Hockney tiled image shut down perception rather than free it? Such architecture of image making has a potential, through place making - placement and allowance, to soar free of any aesthetic law as in the exquisite examples of Braque and Romare Bearden.

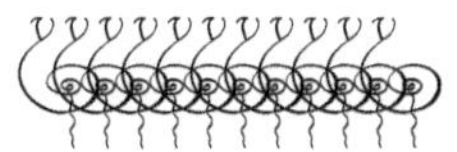

<Narrator's dreamstate notes: the shuttle bus; a small frightened black child trussed in over-sized backpad, expensive clothes, no adult in attendance, stands waiting to get on a shuttle bus, all he remembers are the instructions to sit with *andrew #24* and let him know 'you are his now.' The bus arrives and others push around the boy to get on, the boy is frightened and lost. Scooped up by suddenly appearing narrator, the two present themselves, at the door, to the shuttle driver and a chorus of "what?" ensues. The shuttle drives off with the door open and the two trying to load on. They fail. Boy gets put down, eyes ahead, ramrod still.

The samaritan-narrator says, claiming authority which can never be his, 'wait here, I will be right back.' In his brief absence, a new crowd of the anxious fills in the spaces around the boy while a black van pulls up, back door slowly slides open, hesitates, and closes. From the driver's side a tall black man, dressed all in black, with a black ball cap carrying the number 24 comes round the front, the crowd separates as he bends down and picks up the small one and puts him into the front seat. The small one tries to talk. The tall one smiles, nods and points to the number on his cap.

When the samaritan-narrator gets back with a police officer, the black van is pulling away having left all those others standing on the curb. The narrator says to himself, "I can't use this. It has a will of its own." Only to look up at the arrival of his Enno and Resumé, standing at the curb. "They don't belong here, not in this story anyway." The black van reappears, door opens and only Duden and the dog are permitted inside. It leaves. The narrator clearly sees the small black boy pet the dog. The black chauffeur sends his face back to the narrator with a clear message. "Figure it out, this story is bigger than you, all you may be able to do is pretend you had a hand in it."

"How can I do this story if I don't know everything that happens to my characters? To permit them private lives is to deny the sanctity of the narrator. I thought I had the final say?">

(Answering the need for some semblance of understandable reality, the author allows the narrator to continue:)

The town of Mascara once was a logical way-stop on the difficult highway over the Cascade mountains. A bladder drain and sprain gain before continuing into the deceptive wastes of the high-desert great-basin. Mascara, once a store, a gas station, a two story western-style saloon cum hotel called Hotel, and the storied one-story Speedy Gonzales Motel for unraveling traveling salesmen. Plus, out back, the creepy, dingy, mythic Bitterbrush extended-stay and one-hour lay-over Motor Court, never with any vacancies because it was permanent home to its owners, nine aging prostitutes and their small dogs.

But all of that was way back before the town changed its name in preparation for two golf course developments. In those early years it was known as *Facepaint.* Later the name would be changed to *Cascara,* but that only lasted two days because the wives of the fourteen contractors and the husbands of the thirty-seven realtors didn't like it. So, compromise resulted in changing the name to *Mascara.* Fitting, as this town spent the better part of a century pretending to be western, even codifying the requirement that all buildings looked western. Cosmetic. Some might say phony.

Today there are tens of thousands of residents all mostly concerned about appearing to be happy, well-balanced, top shelf folk. Today Mascara is no longer dingy, or mythic, or obviously creepy. No, today Mascara, the town, is full of itself and alive with its active suspicions of everyone and every thing; today, successfully pretentious, homogenized, sterilized and set well away from the the ugly vulgarities of larger society. Rather than obviously so Mascara is secretly creepy. Thing is, most people know it. Doubtless, with the coming calamities, it is destined to be a difficult to explain ghost town, and ironically, therein finally an authentic western town.

Jackson drove out to Riven's cabin to try to find Enno. He wanted to talk to him about the team of mules and invite him to go to a farm liquidation auction. Now with Titus gone and Jackson's young Belgian team learning fast, he thought the boy might be interested in the old mules, Titus's favorites. And Jackson was more than a little excited to coach him at the auction on harrows and miscellaneous farmstead goodies. Jackson had noticed how Enno's eyes poured over the junk piles and implement lineups whenever he visited.

Always welcome at the Fork Holding (that's what they called their canyon farm) Enno first fell in love with the farm and then came to love the entire Fork family. The big black farmer and the young white orphan carpenter drank strong coffee and talked several long evenings first about their memories of Titus and then about the tools Duden would need to start his homestead farm. Jackson Fork wanted to help and suggested that, in trade for Enno's lending a hand, he might have some old implements that would work.

It had been unusual and heartening for Jackson to find he shared so many things in common with this intense and single-minded young man. Neither of them used cell phones, or computers, both of them were passionate about independence and their constant striving for self sufficiency. Both of them knew they could build and make what they needed. Jackson from a lifetime of experience, and Enno from deep down inside himself.

Jackson had to smile when he learned of the tool trailer. It was an old dilapidated contractor's office on wheels. Enno had proudly taken him inside on a tour of its contents, anxious to hear what Jackson thought he was lacking. In this simple tally they came to respect one another, to know that, with each to the other, each with the other, there were needs to appreciate.

Enno had the rudiments of a carpenter's kit; hand tools, power saws, drills, sanders, squares, levels, chalk line, pry bars, screw drivers, chisels, punches, pliers, wrenches and sockets. And there were lots of sundries like ropes, chains, and such. But Jackson saw no shackles, no hitch pins, no eveners, no neckyokes. He had learned from Sloop, Lloyd and Titus that Enno was keen to do most of his farmwork with draft animals. So it was time to find the riggin'.

Not seeing Enno there, Jackson took a round magnet with a spring clip on it and stuck it to the outside of the door to the tool trailer. Then, in the clip, he put a message along with the sale bill for the auction.

Lloyd seldom traveled further than he could return in his pickup same day. So his art dealer Cash, had to fly in from New York, which meant catching a commuter connection from Portland to Redmond. They sat in the airport cafe talking and going over contract papers. And they were splitting the proceeds from a big sale.

Cash Apercu had consigned an early Lloyd Shoulder's sculpture, Nixon's Toilet, to the latest Sotheby's contemporary art auction. It had sold for $1.6 million before fees. It had been consigned somewhat anonymously by both Apercu and Shoulders, listing both as co-owners under their LLC.

"With your permission, I want to stage a public sculpture competition between you and Bisquets. You would each have a short period of time to work up large drawings of what you might propose to build for Central Park. The theme is to be *civility*. I have two glass-fronted warehouses that will serve as your studios. They both have foyers which will contain seating for paid observers.

Lloyd stared at him blankly, Cash continued,

"His, I'm guessing, will be some sort of conceptual/stage piece. Yours would doubtless be more traditional. Since I represent you both, I will function as judge. And the jury shall be a combination of Met and MOMA board members, of which I am two. I have two internet corporations competing to sponsor and provide the winning sum."

"He's not worthy. Bisquets is not worthy."

"He's thirty years your junior."

"What is that supposed to point to?"

Waving him off, Cash said, "He's getting a lot of attention these days."

"I'm happy for you. With what you brought me today, I figure I'm set so perhaps it's time for us to cut the ties."

"Shit, don't you go cowboy on me now. Long ago you stopped doing this for the money. And I suppose you know I don't need it either."

Lloyd smiled, looked down, and said, "You got that right on both counts, and like I've said before, I don't want to know any more than I need to. No, Cash, count me out."

"Are you still donating to the South Sudanese victims?"

Lloyd nodded and instantly felt his motives pinched.

"The prize for this contest is $15 million. No commission. That would make a handsome donation. You know, with my connections, I could help you make it doubly beneficial for those women and children."

"You bastard."

They fell silent for a few moments. Cash spoke,

"When we first met I sensed a conflict between who you are and how you live."

"I remember, " answered Lloyd, "you came to the ranch and what you found there bothered you. You said back then 'a reasonable man, an artist, can't live like this.' and I answered 'but I do.' Then you said, 'you are obviously an educated man, it will destroy you to separate yourself from the world that leads.' And do you remember what I said then?"

"Yes, you said 'my way of life is not a choice, it's a path.' "

"Well, things have changed since then. Back in those days I was happy and at peace, and it allowed me to work."

"And now?"

"Now? Now that your world has become an inferno of greed, destruction, and stupidity I know I must keep myself separate, or otherwise I would die tryng to stop that train. A waste, if I believe I have something of value to share for the times yet to come."

Cash finally had what he needed. He offered,

"In 1503 or '04 DaVinci and Michelangelo went head to head publicly in Florence, competing to paint the wall of the Great Council Hall."

Lloyd interrupted, "And as I recall neither won. Michelangelo was called away for the Sistine Chapel. Not to mention Bisquets is no fricking Michelangelo. ... Oh, now I see, the instigator of that contest, and its judge, was Machiavelli. That's what you're doing! A proxy war?"

"You come to New York and do this and I will guarantee that money wired to the charity of your choice."

"I haven't won yet."

"Oh, yes you have."

"Listen Cash, I will only do this if it is a legitimate competition. There have to be three judges. You can be one, the secretary general of the UN needs to pick the next one, and the third will be a graffiti artist selected by an NYC beat cop I know."

"It's a deal."

"And there is more.."

Laughing he said, "I recall this being just one thing..."

"Four days and I'm out of there. Plus no spectators."

"Closed circuit camera?"

"Maybe, maybe not."

All during this time, Titus sat on top of the tool trailer and listened in on Jackson's thoughts and Lloyd's wrangle. Somehow he knew he would combine Resumé, Lloyd, Cash and Jackson as a way to continue protecting Enno.

What causes people to spend random amounts of money indiscriminately? Or is the discrimination transparent, even invisible? That second part may be important but forget it for now. Must know the first part. In every man's walleted life there are moments of painless waste, perhaps charitable but more often in the service of self-gratification. Frequently that waste comes when convenience meets haste - no time to quibble, no time to examine, trust the instinct that this is the right thing to do. Few ever keep a scoresheet on the accuracies of an individual's instincts - the variance in applicable accuracies, from one to another human being, offers up the entire soldiered backgrounding of human mythology - a complex web of rationales.

Start with the basic understanding of man's weakness, as it is central to what drives a select few to acquire the suggestive evidence of a higher sensibility, a communion as it were with the intellect in handshake with the spir-it. This is often manifest in the "collector's" mindset. We collect for what purpose? We collect to be able to prove at some point that we are more than we seem. We collect to throw off the bloodhounds of the common good. There is a bridge to be built between the common good and the greatness of the great. But neither the commoner nor the great ones are party to the architecture. The great ones are too self-absorbed and the commoners aren't interested in the view from atop human achievement. What they covet is the view on the way up.

You make a wager, you bet that you can identify greatness before another might, or early enough in the process that the odds are against you. Yes the odds are against you, for therein lies the payoff. Thirty to one against your horse winning. It wins - you win, big. Five hundred to one says you cannot pick a worthy painter before anyone else. You pick one, she wins, and you win, big. More than money, much more because now you feel an earned membership in the pantheon of those few who were there at the beginning. There are two names, ultimately affixed to that museum-hung Van Gogh painting, the painter's and Walter Annenburg's as collec-tor/donor. Some would argue that Annenburg helped to make Van Gogh who he is today. Others, some fellow great ones, spit and piss that Wally brought nothing to the game but money. And so we come to the fork in the road; who, we must ask, told Wally that the Van Gogh was a worthy invest-ment? He did not come up with the notion all on his own, because he was an industrialist of no creative or aesthetic intelligence. When he adores his collection it is with a clear mind's view of the thousands who envy him and covet his objects. He is incapable of trusting his instincts - afterall they have only worked for him in business and this is certainly NOT business.

All of this flows in the brain blood of Cash Apercu, art historian turned consultant bathed in the commissions of contrived auction sales. Cash Apercu, who once used the profit from arms sales to gain entry into the salons and underwear of the art world and who now has successfully re-versed the process. It is art profits which allow him his high stakes pleasure with arms sales. He collects collectors, dealers, distributors, and deals. For him the ultimate pleasure of Breughel's Paradise, hung in his private vault, is equal to and no greater than controlling private security systems in the groin-torn middle east. Yet different, dangerously different because for one he would die and for the other he would kill. And, yes, they flip over occasionally, one to the other. It is Apercu who owns a piece of every new art museum built on the planet over these last twenty-five years. And no one has figured it out yet.

Shirley thought about what Monk had said. "Most people are hunted. Then there are those who are hunters. If you are hunted, if you run and hide in fear, you will be caught or stopped. If on the other hand you are one of the hunters, and you stay 'on your game,' you will never be caught. If you find yourself hunted, you hunt right back, you hunt those hunting you. It becomes a double negative divided by zero. You might be stopped but you will never be caught. Vicky is learning to be a hunter. You and I are most definitely hunters. I by training, you by insistence."

"That's too simplistic."

"Is it? You're thinking about young Enno? Well, I would agree he is in a third group, the very, very small percentage of people who are neither hunted or hunter. Those who are ambivilous or charmed or both. I don't see Enno as necessarily ambivilous. Charmed? Yes, I'd have to say he is. Enno impressed me as someone who makes most evil disappear by ignoring it, that's why I am drawn to him."

Shirley mused. Yes, Enno was something apart. But, if she allowed herself to think it, she sensed that at Duden's deepest recesses, under all that self control, he held off malice, a need to crack heads together.

"Still say that's too simplistic."

"Simplistic? Yes it is, but that is why it is so useful. Force all circumstances to fit and juggle the ones that don't. Otherwise it's mud out there."

"No, I was talking about Enno. He's more than he seems." She regretted both thinking it and saying it.

craquelatured empty
bread rub and embellish

Chapter Four

work

*"Instead of the people owning the money,
money owns the people." - Lewis Lapham*

Squash Blossom Soup
(Sopa de Flor de Calabaza)
*Good luck finding a specialty market which offers canned
zucchini squash blossoms. And if you do find them, make sure
you cook them immediately after opening the can. This soup
features 2 lbs of blossoms, 9 tablespoons of butter! 3 green
onions sliced, 6 slivered garlic cloves, 6 chiles serranos, very
thin slices please, marjoram, parsley, thyme and tarragon, salt
and fresh ground pepper, 8 cups of chicken stock, 1 bone-
less chicken breast (or grouse if you've the taste) cubed, sliced
mushrooms, and 2 cups of thick cream!*

*Remove the stems and pistils from the blossoms. Rinse
gently and coarsely chop. Melt the butter, add onions, garlic
and chiles. Sauté with blossoms, stir in herbs. Heat chicken
stock separately. Sauté chicken (or grouse) separately. Sauté
mushrooms separately. Mix together, season to taste and blend
in cream. Eat this alone or with someone who epitomizes
abandon. Wow!*

*Serve this soup with a chilled glass of Lois, that Austrian
wine, and a chunk of hearth-style poppy seed sour dough*

*bread. Wait 7 minutes, after eating and drinking, to consume
an espresso and raspberry orchestrated chocolate truffle. Then
jump up and pretend there is an emergency requiring instant
action, like, for instance, a fire in your house. Run around
frantically trying to focus on what should be done. Then fall
back into a soft chair laughing and pick up this manuscript
where you left off...*

*"I think therefore I am is the statement of an intellectual who
underrates toothaches."*   *- Milan Kundera*

Enno drove his old Chevy pickup away from the parts house and towards the weigh station before town where Benji Aboo had left his stranded Indian motorcycle. Couldn't see it at first. It was parked behind the billboard, beside the weighmaster's shed. Benji got out. Resumé, still shook up by the bullet passage, laid down on the truck seat and trembled. Enno Duden got out slow and hesitantly followed the little black man. The motorcycle was fifty years old, cut low and comfortable for Benji and painted in camouflage colors, these over some sort of refractive material which morphed colors and shapes as eyes moved over and round them. The light just so and you would have sworn the bike was orange, another angle and it was solid blue. There was a sidecar with a leather lid. Benji opened it and retrieved a small tool box and an envelope. He handed the envelope to Enno.

It was a letter from Shirley.

"Now it is clear to me how much I love you. You don't want to hear it but we, you and I, are at great risk. But please don't worry, you must trust me. They are learning who and what you are - we are - and they know they must stop us. I will have my other answers soon and then I will come back. For now please

trust Benji as you might trust me. Soon enough, much too soon, the snarl will come to confuse. But that won't stop us. I want to help you build that farm, and for those who might care, we will give off light, as will our children."

A dumb letter. Extreme, milky, feminine though it sounded like it came from a male. Offensively controlling. No signature. To this his first read, it had sweetness and sincerity that inflated the answers he wanted to read. His questions had been gaseous and now here he had gaseous answers that any mature attorney would find non-binding. Enno sat back hard, expecting a chair or something to catch him, but nothing was there. He landed on his tail bone and rolled slightly before bouncing, awkwardly, back up to his feet. Benji watched and smiled.

It had happened so fast. She and he falling into comforts with each other. Comforts and then hungers quickly followed by worries and denial. And all of that in the midst of the volcanic eruptions and Titus's heroic death. So, in the throes of loss and confusion, when she had announced she had to go away right then to tend to her personal business, he took that as clear evidence that they were not to be together.

Odd, his long orphaned heart had never permitted even a glimmer of sharing his life's energies with anyone except his new companion, the old dog. Not until Shirley seemed to fall out of a tree right on his head. The images of the two of them building a chicken house together filled him with the giddiest of joys. The opposing sight of her preaching at him about his significance and the pervasive needs of a crazy outside world gave him a headache. So what. It was clear that he was also in love. And that, beyond his understanding, would certainly muddy everything up even further.

(Back, growing up, decades ago, the lacings which hid the unmentionables were the emotional drivers that filled the hormonal chambers of rapidly growing hearts. We were young then. Being in heat without a clue as to what that meant, we were nevertheless governed by a great and genuine fear of humiliation. That is in part what made Groucho Marx a hero; he floated above humiliation, consummate - goaty - wise fool that he was. Speaking of heroes; for most of us a mentor, example, model or hero was only as large as the hole we needed to fill. Oft times we abused our experience and memories of those people we projected towards. How frequent it was that the ones we signed on with were the ones whose bearing stood out. It wasn't about goodness, or actual strength, or skill, or deed. It was about bearing and balance; how that person stood there in front of us; walking towards us, walking away, waiting, moving in the work. All the cornpone we assign and assigned to this mentor business is just that, cornpone. Because you see, with time we found that those giants from our past weren't so big after all. Sometimes we learned they were outright liars and frauds, lovely ones but fraudulent nonetheless. They 'seemed' to hold us to standards they themselves could not meet. And why not? Isn't that as it should be?

Back to that humiliation note: today abasement and degradation have taken turns in odd directions. It still works for some but it is emotionally proprietary in ways only an old tattoo artist can fully understand.

Another digression, another wrinkled dollar from the pocket of the twice washed jeans...)

Houston Tuttle sold ideas. Not objects but ideas. He was so good at it that it meant balancing the night-side of his brain with the day-side. To maintain his fragile authenticity he cobbled up a chambered home-base as anchor. This was a place to return to when the work was done. This was a place to hone his truculent sense of purpose. He was imperious when zoning in on a target, imperious because most days he sold even himself the whole damned lie. Did it so well it informed how he dressed, how he leaned into his walk, the comforting cloaking smile which waited to be called up as weapon and closing devise. The illusion was that he knew what made the world tick, and for a fee he would come stand before your group and give you all of his secrets. His subject? The motivations of motivation.

He had started as a salesman, selling sales opportunities. In the beginning he was entertaining and vapid, funny and staid, sexy and flatulent. With time he became as sticky-smooth as molasses on a warm baby's bottom. Soon he was being asked to 'speak' rather than to sell. And quickly he realized he was the product, he was selling himself.

After the one rotary club talk a woman had come up and asked what he would charge to speak to her Unitarian church group. As she shook her head yes, he said that to come back to this town and talk would require a couple of thousand dollars and all his travel and accommodations covered. He offered this as a polite reason why it just wasn't feasible. She responded, I think we can cover that. When can you come, tomorrow maybe? And, I'm sure that we wouldn't be interested in the sales stuff, but that other stuff you talked about - that stuff about why we should try to be better. That's the stuff they will want to hear.

The next evening he pocketed twenty five hundred dollars and smiled about how much all of those nice people appreciated his stories, his words, his manner at that lectern. That's how it got started. Houston Tuttle got ever higher fees to just stand before different groups and talk about the 'whys' of life. He was his own man. No ties, no contracts or obligations. Several nice bank accounts. A phony address for legitimacy. Until she arrived.

It was early December in Manhattan, he was to speak at the Red Lion to a group of retired school administrators trying to organize themselves into a force as a volunteer organization. He had scored the astounding fee of twenty-one thousand dollars for the evening's presentation. Astounding for someone who had no offstage life, no published books, no company affiliations, no degrees, no institutional creds, no family connections. He was a blank slate OTHER than the fact that every time he spoke he sold himself into yet another venue. Tonite would be a springloaded sale. Tonite the cement would stiffen around Tuttle the Legend.

Bessie Quince, clawing her way for months through New York City's art education underbelly as she kept a bead on her goal of absolute power in the arts, was at this evening's doings to gather a feather from the wealthy woman who frothed the group's futurings. When Houston went on stage, following his bare bones introduction, Bessie found herself transfixed.

His movements were slow, sure and magically orchestrated. It seemed that he had several hands, one to sweep over his sandy hair, one to hold some note cards, one to pull the microphone from its stand, one to cup the back of his neck as in reassurance, one to hold along the side of the podium, one to reach out and cup the listener's chin, one to do little fingered dance numbers to punctuate the air in anticipation of words yet to come, and so it continued. And the eyes. Thank heavens there were only two. They were grey, backlit by silver blue and rung around by a peach tone.

And Bessie, transfixed - but learning every second of the way - was ready to catch the significance of what happened next. Before Houston Tuttle said one word, a gloved hand with a rolled up program went into the air somewhere along the fourth row. It could have been a request to ask a question, it could have been an auction bid. He saw the hand go up and, staying very still, hands flat on the podium, he nodded but only with his eyes. To which the woman said, "I will." A moment of silence and then a rush of nervous laughter from the audience.

Bessie saw all of this and understood it because she too felt, as she watched this man, that she wanted to raise her hand, to sign up, to join up, to be acknowledged. And she knew that most in the room felt the same way. She thought about it as she watched and listened to him speak. It wasn't a religious thing, it was a social tonality. It was as if everyone had elected this man to decide in what order they would leave the boat, in what order they would be seated for dinner, he was the bailiff the judge would defer to, he was the town baker, he was the court poet, he was the confessor to kings, he was the father of all bartenders, he was the substitute basketball coach who wins the national championship.

Even before Houston finished speaking, Bessie felt herself move in his direction and as she moved she thought of how it was that Cash Apercu had to meet this man. Ah, but not before she had carefully, surgically inserted her twenty hooks deep in and under the emotional psyche of Mr. Tuttle. Her Mr. Tuttle.

Shift

"I'm going to follow you home," offered Benji to a completely distracted Enno Albert Duden.

"That's where we are going now, to your home. Don't try to come up with answers to all of your questions right now. It can't happen."

Enno nodded and started up his truck.

As Sloop drove by, on his way into town with Helga, he saw the boy, the dog and Benji. For the briefest of moments he allowed himself to worry about what the Duden kid was getting into this time. That sympathy left quickly as he reminded himself of how his world and work had collapsed because of this hapless kid. Helga crippled by a bullet, Titus dead. The vast organization they had built to help poor farmers all ripped to pieces because he, John "Sloop" Ogdensburg, had made the mistake to trust this kid, and his erroneous instinct about the kid. Duden was not the "one." The boy's enigmatic personality came from his simplicity of mind, not from a deep-seeded gift. Sloop had trusted the 'balance' of his life for so long that he was blind to its sudden absence. Shoulders was right in a way. He shouldn't be blaming the kid, but it all seemed to start there. Innocently enough, he had thought young Duden would continue as an apprentice to his own ranching, that over time Sloop could hand off much of the labor and enjoy being the overseer. And that as the chosen "one," he, *Digger* Duden, would slowly come up to speed on the organization. As John was thinking these thoughts, a black and white Titus was able to slip in and nudge Sloop's periphery, just enough so that he would notice, over there, that other man with the marine bearing and the aluminum case as he stepped into the old '79 Ford pickup. Something not right about the picture.

Jeb caught Sloop's glance and all his Baghdad training told him he had been made. Was it the truck? The case? Him? His clothes. It was the truck, now he could see it. When he got in to Redmond he had to decide between the Black Escalade he was so familiar with or the older pickup. The Escalade was an easy rent. The older truck took some leg work. On the surface it seemed the truck would help him to blend in. Now he realized that in

Mascara, this lazy pretentious resort town, the Escalade was the invisible vehicle. Especially with a little dust. He'd soon fix that.

The hair stood up on Sloop's neck. He had allowed himself to believe that the murderers had all been gathered up or lost in the riots swirling during and around the earthquake and eruptions. He thought of himself and Helga. And he thought of Duden. He had to get to the boy and warn him.

Dr. Fil's anonymous blog post:

Computer graphics programming has shown me the answer. Up 'til today, we, all of us compliant human insects, have seen the world, our world, the regions, the cities, the nations all as delineated by somewhat arbitrary mappable three dimensional boundaries. We can write each up in a set of quadrants and register it all as so many surveyable "mining" claims - for in essence that is what each and every one have been. "I, we, hereby register that, from the stake over there to that one and that one and so forth, I, and only I, have have have have the right to strip whatever we deem of ultimate value to us me us me us me us me." (Multiple subjects delineating corporate insertion - share the blame - share the profit - share the shame).

We require new thinking, which applies new science to shifting paradigms, in other words; we're screwed if we can't figure out what to do next. It is said we can't figure out what to do next unless we understand where we are, where we are heading, and what the word <u>perspective</u> means. We've made a "Wreck" of things. Therein lay all the answers.

If we chart the outermost points of human influence, going even out to unmanned space flight, and make of that an expanded envelope-graph inside of which resides the planet earth, we have defined a universe. Some might begin to see it as a footprint, I see it as a stink print.

I have registered, in the public domain, a claim, a copyright, a trademark (all film rights reserved separately) on that territory by patenting it as a new thought form - an essential territorial domain, a measurable sphere of influence, a requisite testing platform for the creation of new and beneficial thought. I have registered this domain as "The Wreck of Human Arrogance" or for short

"Wreck."

(Here's where I need a song, Gordon Lightfoot singing 'The Wreck of Human Arrogance.' A long song, a fado-infused song, an African wail of a song, a Keb Mo, wet-stretched-socks-in-the-trees, *song, a Canadien donut and black hot water song.)*

*From this day forward, I say that the human state of influence we now know as "**Wreck**" has dominion over any and all man-made bodies of governance and that most certainly includes nation-states, organized religion, corporations, the management companies of rock and roll bands, social networking, internet shopping and all mainstream news organizations. (Have to do that last one so we can fire them.)*

***Wreck** will issue a new form of currency as the standard of exchange. It shall be called the **dick** and it is to be guaranteed by liquid ounces of the new element **Filenium**. As a starting point five thousand dick will equal one ounce of Filenium. The flag of **Wreck** will contain an image of the 'blodget,' that soon to be universal symbol for Filenium sometimes mistaken for a blonde widget. The seal of **Wreck** will swirl around a big garbage can which has a sad face painted on its side.*

***Wreck** will immediately outlaw any human or human-engendered endeavor deemed as adding to the Wreck's circumference, weight, stench, biliousness, entertainment news, temperature, or stupid jokes. Conversely **Wreck** will encourage all activities which increase bio-diversity, limit population growth, increase fertility, encourage intelligent humor, expand on creativity and feed everybody.*

You might reasonably ask, how you gonna do that? Answer: by providing access to a heretofore unknown basic element, Filenium, which has properties that will revolutionize energy, transportation, industrial process, human relations, and representative government. I've got the Filenium, I'm going to show the world what it can do and I'm going to offer it to the public domain IF the governing body I have just described is set into law. This is not a threat, nor is it blackmail. It is a sincere offer backed by damn fine science.

Dr. Fil's finger poised over the *publish* button for just a second and then he withdrew it. He was nowhere near ready. They would laugh him off the planet. This plan was too good to be trivialized. He wasn't up to the swagger and machinations it would require but he knew how this had to proceed. He had to make a dramatic presentation to a handful of highly prosperous and greedy nation states Japan, China, Germany, Brazil, Russia, France and Switzerland. And he had to keep it out of the hands of Apple, Alphabet, Amazon, Facebook and Bayer.

Before those moves and protections, he had to start the cogs in motion one territory at a time. Oregon would be first. It was a natural with 56% of its land mass in public ownership and being the secret wellspring of so much Filenium.

Dr. Fil had made inquiries to try to find who should be selected to make the first overtures to the federal government. He was sitting on trillions of dollars and he was of a mind to trade that for private ownership of the Oregon public lands along with the release of the state to its own sovereignty. National debt what it is, this one time payment would make the remaining United States debt free. But the plan required an agreement with a country, or consortium of countries, that could secretly finance such a purchase, finance it for a piece of the land and a piece of the future distribution of Filenium.

And Dr. Fil knew he wasn't the one to make the overtures, he needed someone like Dr. LaLouche, the ex-Utah senator and successful Mormon marriage counselor. LaLouche was folksy, crass, handsome, disreputable, dubious, well-oiled, cross-eyed and a claimant of spurious French heritage. He was perfect to carry talk of obscene amounts of money to the high humidity denizens of the nation's rancid capital while texting the three people in control of the European Union.

With the sovereign country of Oregon mining and refining Filenium for a hungry, cold and anxious planet it would be a matter of a few years before Dr. Fil could make his plan for **Wreck** publicly known. At that point no one would laugh. No, they wouldn't laugh, but they would load their guns.

She was trying to decide which cheese to add to the snack tray she was planning for the party. She read the labels and found herself traveling in her imagination. Eyes closed she saw a rocky, semi-barren landscape.

Roquefort, from the caves of Combalou in southern France and felt by most to be the world's supreme blue, the recipe and name protected since 1411. This heavenly flavor reminds one of the cavern air where the cheese ripens and develops over a three month period. Try this cheese with a glass of Sauterne and feel the perfectly balanced creamy chunks dissolve on the palate like sharp, soothing, milky lozenges.

Taleggio: Beginning in the ninth century, cheesemakers in Northern Italy masterminded this washed-rind cheese from cows milk and aged it for more than 35 days. A meaty, nutty, almost salty wonder with fruit floating through. Aromatic. Serve with Riesling.

She could see the rapid rise of the Dolomites.

L'Etivaz of Switzerland

She could feel the mosquito bites.

Butter Kase of Wisconsin

Now, she had to find just the right thin, bean-flavored crackers.

Perhaps Madagascar.

Resumé stood still, lips pulled back, and allowed Enno, with a thin little piece of kindling wood, to pick the irritating bits of old bubble gum from between his gapped front teeth. Enno chuckled.

Way up above Titus frowned, unable to understand what was funny. Come to think of it, ghosts are what's funny, most of the time the joke's on them.

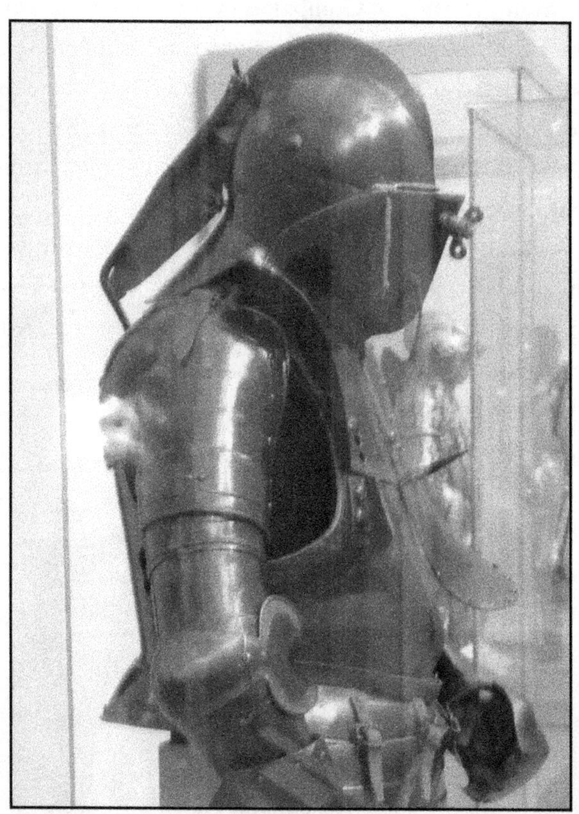

Chapter Five

the magic

__ensellure:__ the concave curve formed by the spine:
in a woman, the lumbar incurvation.
"...thus to map so small a corner of so great a world."

"A frivolous society can acquire dramatic significance only
through what its frivolity destroys. Its tragic implications lie
in its power of debasing people and ideas." - Edith Wharton.

It is afterall, the mind exhausted by the tedium of anxieties
which waits most patiently to release itself
to the greater designs. True energy, measurable energy.

"It is quite simple." The triple-breasted Russian business man offered.
"If I fail to end up with both the Rauschenburg and the Shoulders you are
dead." Square, balding, jaw locking in a straight line as frame for his thin
lips, doughy Anatoly Grim spared no expense on his clothes, toilet, and
personal hygiene and yet he still came across as a large rotting potato.

"Anatoly, it is no wonder you frighten people." Cash smiled as he spoke
and you could see the sugar of this handsome man's mouth freeze and crack
when he added in his silky French accent. "Listen you toad, your heart has
stopped beating. You are already dead, any chance that you might rejoin
the living, and in triumph, is up to me. Of course I want you to succeed,
for the nasty soup I would make of your living kidneys will challenge my

chef's magic."

Grim blurted out a sackful of laughter. "Apercu, damn it, you must be Russian, I insist on it." And the jaw set once again as he added, "For I want you as my brother when I kill you. Nothing less will do."

"Good, then I have some time. Now, please listen, you have an important part to play in all of this and it is silence. I understand that you want to be at the auction, seen and envied. But if our plan is to work you must divulge NOTHING of the arrangements. Is that clear?" He calmly and reassuredly imagined a point in the future when Grim would be dead. And as counterpoint, Apercu had a flash auditory vision of the Castrato Arias he so loved.

and then, over there...

Resumé, the old stock dog. tried to sleep on the corner of the bed. His head was tired. At least it was now apparent to him that someone, someone invisible, was getting inside there and talking with him. It was also clear that these commands were not to be followed. Listen. But do not go there. So he trained his thoughts on memories of Rita and his gonads took over.

Benji had seen enough to know.

"Yes, for a long time now there have been dates and places charged with electricity. Points where that spirit stuff, as you call it, slams together, converges. No coincidence that we've been having all these extreme storms, hurricanes, earthquakes. It's not just climate change doing it, it is larger theater at play. Convergences, there's a big one coming up. It's about to occur once again, maybe the most important in over a thousand years. And this time it is not to be directed at a place or time but rather at a person." Pause. "Seems you may be that person."

Benji Aboo was sitting cross-legged on the floor of Sam Riven's hideout cabin, the temporary home of Enno "Digger" Duden. Enno looked at Benji blankly, his expression did not change. Benji continued to speak.

Then Duden started to slowly shake his head no.

The small Aborigine looked at the boy as though he were a landscape about to morph, as though he were a mirage about to change shape from long and thin to tall and short, as though he were a pond's surface waiting beneath black silvered clouds for those first raindrops, as though he were a catbox that needed cleaning.

Benji held Enno still with his eyes and said,

"In the language of ancient Gaul it is the pourquoi of humanity, the why of men, it is the promise, it is the empty beginning full to overflowing with answers, there are no previous lives contained within its essence, only clues, sacred chemical stains that propel it towards ..."

Enno shook his head muttering a string of no's. Benji caught his eyes and calmed him.

"In our aboriginal dream-stories, if it is you, you are the beautiful frustration, in Tinga Tinga stories you are the one to be blamed for everything and the one to be treasured. You are the missing back story.

"All of the rest of us are shaped and limited terribly and gloriously by our previous incarnations. You have no previous incarnation. You are the origin returned as the control against which the failing experiment of humanity will be measured.

"If it is you, you will tell us nothing and we will tell you everything and our energies will pass unrequited. It is yet to be written how you are allowed in our midst, whether your story is to be amplified any time soon, for it will be amplified eventually and forever."

Curious how Enno, rejecting these observations of Benji's, still found in them no threat, only a version of reality he thought he might ignore.

"She thinks that too. Shirley, I mean. No, maybe not exactly that, but something that feels like that. But then I see her looking at me other ways too. It's like she sees me two ways. She has some similar ideas."

Enno drops his head. "But it's not true. I would know." He raised his head to confront Benji's eyes, "wouldn't I?"

Benji said, "Decades ago, when I was twenty and counting, old men, those over 35, were stubborn, mean, wise, solid, emotional, grand, wild and thickening. They were predatory, beautiful, lazy, masterful, sarcastic and dismissive. They, many of them, knew how to work. They understood treasure, loss and nastiness. They often cradled greed and depression. They feared beauty and getting caught and not much else. They were championship, steadfast and thieving. They were, as I watched, rapidly becoming the shape and texture of their lives, that their mother's father's lives had bequeathed. All of that and more, when you lumped them together. They gave direction, some of it golden. But one thing ran common, when I was young I was certain those older and wiser just didn't understand. That has always been one of the important balancing tools, contradictions between observed and felt.

"Now, today, we, you young - I old, have this in common; you, in the eyes of the many, are guilty for no other reason than you are stand-offish, you abstain from society. I am guilty because my physical appearance and manner threatens society.

"And you attract people who stand outside of society that they may continue their work. We are a society without."

Enno interrupted, "All of you are thinking too much of me. I don't deliberately avoid people. Well, maybe I do. But all of this stuff, it isn't me, I don't have it. I don't have any secrets. Can't you see? I don't even have the education to know what you are talking about. I'm just a carpenter's helper who wants to have a farm of his own. I've got no family, no past that goes back before anything I can remember. This kind of talk tires me out and it worries me, more and more every day."

"Why now more than ever?" Aboo asks.

"Because, you read her letter, she says she's coming back but only because she thinks I'm that one you all keep talking about. I want her back, yes, you bet I do. I want her back real bad. But I can't pretend to be what she thinks I need to be. I don't have what it takes. I can't act. I want a farm, that's it. Nothing else matters."

One word the young man said caught Benji's attention sideways.

"None of this has much of anything to do with you and she getting together. Whether or not you do won't affect what's coming. But then, truly no one knows for sure if even that is right. Maybe it is about you two getting together. In any case, Enno, I am one who can tell you that the girl is in love with you. And the only demands being made on you by her, and they are unreasonable - yes, are that you stay alive and wait for her. With all of this, with any of it, being yourself is the key. But I must ask you something, back there, you mentioned 'the one you ALL keep talking about'? Who do you mean by ALL?"

Enno heard him say 'she's in love with you' and he shook his head wide in disbelief and thirst. It took moments before he could answer the question.

"Oh everybody: Sloop, Jackson and Brenda, Titus when he was alive, Sig and Vicky, Nettie - not so much Sam - and a bunch of others I don't know. Oh, and that Vic at the feedstore..."

Out of all the names the only one Shirley had not mentioned was Vic. "Tell me about this Vic."

"Vic Nib? Oh, he showed up a couple of weeks ago and went to work for the local feedstore. Nice enough guy. Nosey. Always asking questions. Why?"

"Describe him to me please."

"Oh, five foot eight, silver blond hair, moustache. Kind of reminds me of a track coach I once had. Nothing unusual about him except that he wears just one earring."

"One earring?"

"Yes, the other ear has a jagged notch where the lobe should be, kind of like it's been ripped off. Come to think of it, anytime Vic sees you looking at that ear he turns his head sideways to hide it."

"Anything else that stands out about him?"

"Oh, he's pretty cocky about his pain thing. Likes to strike a match and rotate it into his arm to put it out."

"What's he talk about?"

"He's full of wacky opinions about what's wrong with the country and he mixes that up with big words about philosophy and stuff. He keeps saying that he thinks I'm destined for big things. But he says it like he wants me to respond. Why are you interested in him."

"He, son, we have to practise staying away from."

"There's one other. A new man around here, a scientist. Calls himself Dr. Fil."

They were startled by a knock on the door.

Benji moved quickly and calmly to stand, back to wall, on hinge side of the door.

Before Enno made it to the door an envelope appeared on the floor. He opened the door while bending over to pick it up. Envelope in hand, still bent over, he looked up surprised to see Sloop get into his truck and drive away slowly. Why was Sloop limping, what was with the rifle?

The note read, "It begins again. Return to the codes."

Outside, in the night, perched in his overcoat on a juniper limb, the ghost of Titus Ibid watched from his black and white world pleased that he had figured out one small acceptable maneuver. Yes, he saw the other man behind the tree watching. That man with the torn ear lobe would come in handy. Titus Ibid put Vic Nib in his ghost pocket as a diversionary stone - a lob lolly - a false bird call - a lint roller - a basket of plastic fruit.

Vic Nib had followed Enno and the little man to this remote cabin. He was very careful not to be seen, or so he thought. Watching the two of them so intently he had neglected to look back. Sloop saw Nib, followed Nib, and understood him for what he had to be. So there were two of them, he thought. He saw the stone outcropping behind which Vic parked his car. Vic scurried off through the pines and Sloop waited a short while before driving his pickup truck on past Vic and towards Enno's temporary home. He pulled the collar of his coat up, took his trademark slouch hat off and mussed his hair. Envelope in one hand, rifle in the other he cocked his head to the side and pretended an exaggerated limp as he openly approached the cabin door, knocked, slid the envelope through and limped off.

Nib saw him drive by - park - get out and approach the cabin. Nib saw the rifle. Nib saw the boy come to the door and the old cripple drive off. Nib knew to follow the old cripple. The license plate was missing and the truck was non-descript. It drove off slow. Nib ran crossways, through the woods, trying to get to his car in time to follow the pickup. He needed to know who this was.

Titus floated above it all fighting the urge to smile.

At the car Nib never saw the waiting form of Sloop, had no way to expect the handkerchief of ether over his face, didn't feel his wrists being tied behind him and to his ankles, didn't feel those eyes looking at everything in his pockets and in the glovebox, didn't feel his shirt being opened to expose his chest. Sloop had been ready for this moment. He took out the little packet, lit the red candle and let the hot wax drip to the man's lowerchest. He then took the match stick and drew an impression of a trident in the wax. It should have amused him but it didn't. The diversion had been made available, 'nuff said, more than enuff thought.

Next morning Oregon State cow-cop Fred English would find a parcel on his front porch. It would contain a pair of tennis shoes, a wallet with ID, and a note which read "There are two of them in town again. I suspect they don't know each other. Here's the location of one. Same storm - button hook. If you leave him more than a couple of days he will be dead. Your choice. Monk."

Long ways away but close in spirit...

Shirley had just one thing left to do, then she could return to Enno. She finally had her answer. She learned that her mother had drowned in a boating accident trying to cross the Columbia River bar at night during a storm. Her crazy mother had joined a group of fishing widows determined to board a Japanese factory-fishing trawler in protest. They had never made it. But before they left they had spent the night together at a campground called Dismal Nitch. A nasty soggy cove aptly named by Lewis or Clark or one of that troupe. It was said that all the protesting women, understanding the danger they faced had put together a mini-time capsule; a container buried at Dismal Nitch, with messages to the future, maybe even messages to those left behind. Shirley thought it silly but she had to check it out. For her, closure was an absolute passion. She was satisfied that she had the answers she sought about her mother but the whole road had to be traveled.

It was not lost on her that a few months ago, at another way-station along the Columbia, she had unearthed D. B. Cooper's long lost, hijacked funds. And that it had then taken her somehow straight to Enno.

One thing she did look forward to, a chance to find some chanterelle mushrooms out under the firs and rhododendrons. She would ice them and hurry back to fry them in butter for him. At no other point in her life had she felt so completely consumed by her future, consumed and scared silly. "Please don't let anything happen to him!" she prayed over and over again. She could still feel the warmth of his palm on the small of her back.

The public radio broadcast was called "New Convergence." Every week a different, over-educated ex-pat of the sixties brought new sonic and ionic, dietetic blasts to liberal geriatrics. Most of the time it was feel good about feeling bad, or understanding your fears and learning to massage them. This week was only slightly different. This week instead of discussions of expanded consciousness or group colonoscopies it was about new demons fast approaching.

"So, Jerry, I know you've thought long and hard about this. Simply put who are the new devils?"

"Terry Lee, I know you and your listeners will find this hard to believe but it isn't Bayer or the Gee Whiz Corp any longer. This lingering recession is killing them off. And those nasty bioengineering firms? They are fading to become nothing more than backgrounding for the new big brother-hood. No, the new demons are all the strongest darlings of the internet; Microsoft, Apple, Facebook, Google and Amazon, they're the ones who are gobbling up absolute control of the consuming public. They are, between them, far ahead of all other industry and government information gathering systems and well into behavior modification for the public. They will soon help you to decide what you want to buy and keep absolute track of the transactions. What follows is to have those computers then anticipate your every wish with just enough success that you will comfortably forfeit even thinking about such things in the future."

"These aren't such new ideas. The Orwellian threat has been hanging out there in the ether for decades."

"Yes, but the difference now is that control of our minds is a given, to the extent that they see us as pawns in this cosmic chess game they are playing with each other. What we must fear, above all else, is the inevitability of consolidation."

"Consolidation? What has that to do with..."

"Terry Lee, As long as the CEOs are working feverishly to outwit one another, as long as they are fixed on 'acquiring' all of the competition, we have a little time. BUT, when one of them wins and controls all the other information companies, we are faced with certain doom. For you see, whether it is Steve, or Jeff, or Elon, or Bill or Rupert's grandson, it doesn't matter - what matters is that that person, in momentary triumph, will be all alone with himself. Emptiness is all he will have. No purpose will remain. The bubble of false magic will burst and those of us who depend on the system to run our lives will self destruct as the system self-destructs."

"Steve? Do you mean Steve Jobs? He's dead."

"Terry Lee, you can't honestly believe that. A man in his absolute prime, controlling massive amounts of money and technology; a man who we know is insanely protective of his private persona; a man who with his phobic behavior makes Howard Hughes look like Donald Trump; can't you see - he's faked his death to give himself the best position in this chess game I speak of."

"Wow, Jerry. That IS 'out there'."

"Thanks Terry Lee, I know it's entertaining to speculate on these things but I really am serious about this. That is why I am joining the Concerned Representatives of Allied Purposefulness, CRAP for short, in their hacker supported efforts to create penetrating fogs to permeate the data bases of the world. Fogs that will ooze into those 'cloud' systems and act like electronic mirages distorting data. We're on the verge of discovering how we give these fogs a synthetic fertility allowing that they breed and send their offspring out beyond and into the limitless reaches of cyberspace. Terry Lee, what we are talking about here is a program to destroy the digital world through a synthetic fourth dimensional venereal scourge."

"But Jerry, that would mean that, that, like Amazon wouldn't work any more and that like Twittering would be gone."

"That's right, dear lady."

"But Jerry, that would be horrible, what would we do without all that information just floating out there, that information we trust, those hospital records, the email addresses of political donors, the podcasts of the Daily Show? Where would we turn?"

"To books, Terry Lee, to books."

Survival is no longer the first order and that goes against the biological grain. Once it was innovation for survival's sake. Now it is innovation for culmination's sake.

Fred English, self-ordained concierge vigilante, was thrilled to receive the message. Everything else fell away, he understood the instructions. He needed this worthy battle. He gathered up his forensic tools, an electric hot prod used to load cattle, a camera, his shoulder-holstered Beretta and a packet of zip ties. He pinned his badge upside to his ball cap. Then he drove to the spot the 'Monk' (actually Sloop pretending to be the Monk) had indicated. Ahead of him he saw the body of a trussed man thrashing around in a pile of pine needles. Before approaching Vic's sightline, he searched the area and found the pickup truck, the sniper's rifle, a cell phone, a half eaten bag of Fritos, and an envelope of photos. He recognized several. He was surprised to see his photo in there as well.

Arming himself with the electric cattle prod, he stepped into Vic's view, careful to make sure he saw the prod. In his other hand he had a razor knife with which he cut the cords around wrists and ankles.

Released, Vic made a lunge for Fred who calmly pushed him in the chest with the electric prod. Vic jerked spasmodically and fell back to the ground.

"Just lay there and listen. That red-wax embedded seal on your chest should tell you a great deal. If the Monk had wanted you dead it would be done. Instead he wants you to know that the people who hired you to neutralize the target also hired another to make sure you were killed upon completion of your contract. That's two, with Monk that makes three, with me we start multiplying. I'm intrigued. Monk is pissed. He has gone to neutralize your boss unless you do it first. His suggestion is that you go there and tell them the job is done and get the balance of your money. If Monk gets there first, there is no chance of you being paid the balance. Very little chance of you even surviving. You have been under constant surveillance by at least four separate professionals, most of which are eager to prove to you that pain is no reward.

Okay, I'm leaving now, and taking several items for law enforcement. Have a nice day."

Nib reached for his torn lobe and muttered, "I've got you now."

Jeb, the other hit man, drove the freshly rented black Escalade over the forest service gravel road at 40 miles an hour. Satisfied that dust covered every inch of it, he then cruised Mascara's main drive.

Sloop saw him, drove around two blocks and came back down the street from the opposite direction. He looked directly at Jeb and cocked his finger at him and shot the 'I see you' bullet. Jeb said "shit."

the devolution had to start someplace
the crease in the arch braced with
promise gone only worthy question
will the circle complete itself

Chapter Six

patterned landscape

*"...poetry is a comforting piece of fiction set to more or less
lascivious music - a slap on the back in waltz time - a grand
release of longings and repressions to the tune of flutes,
harps, sackbutts, psalteries and the usual strings.*

*As I say, poetry, may be either the one thing or the other -
caressing music or caressing assurance.
It need not necessarily be both."*

- H.L. Mencken

If we fail to wind the watch. If we come upon it static, stopped in time's track, we come upon yet another failed start. To wind the stopped watch, acknowledging our forgetfulness, is not so much to start anew as it is to try again to make of our false continuum a long story, a purposefully long story, one which gathers unto itself. At that point when we realize we have failed, and the only thing to show for our efforts are many false starts, we need to go to the hat check man with stub in hand.

"Return to the codes." That's what the note said. Enno remembered what Titus had shown him, what Sig had tried to help to interpret. Benji went with him in the morning and they scanned the Mascara Market bulletin board. There it was. The vague ad about chickens for sale. Enno used the one last pay phone in all of Mascara and called the number. A meeting was set up with Sloop. Enno to come alone.

Elsewhere, elseways...

He had protected his anonymity. For all those years of providing software tickles to help Titus and Sloop with their *Glass Pirate* wealth redistribution efforts, no one - not even they - actually knew who he was. They had protected themselves from him and he from them. No way to trace back. And now, when everything had hit the skids, Titus gone and Sloop ordering everyone to stand down for the foreseeable future, he found himself happily with time and open thought available.

The slow steady trickle of money from Titus had been good, it kept him oiled just enough to paint, to travel to great museums, to flit around the edges of the art world, but always unknown and unknowable. Always the Hiram Holliday of art spies, a meek professorial ascetic in search of the little secrets. He adored the stories behind the stories. For example, he had long been on the historical trail of who Jules Pascin was supporting when the brilliant painter suspiciously died, some believe by his own hand.

He thought otherwise. The clues led him to believe the Jewish painter had been murdered by a sniveling, egomaniacal coward who feared Pascin's creative fluidity. The clues led him to believe that the little misogynistic Spaniard had set it up. Picasso could not stand the fact that every prostitute in Paris owed something to the handsome little Jew. Pascin was loved all about, while Picasso, the creepy little muscle-bound bastard and sexual predator, was a poisonous thorn in everyone's side - feared and revered all at once. It mattered in curious and destructive ways that Pablo was a genius to be envied; because for him there could never be enough adoration. He consumed and then destroyed the women in his path as well as other artists. Modigliani, Derain, Braque and company were pulled in then stabbed repeatedly while drowning in the urinated public abuse of Picasso's feminine gossip. Only Matisse escaped the Spaniard's bitchiness. And that because Henri saw the rival as no rival at all, saw him that way because Pablo's insatiable thirst for absolute supremacy kept the Castillian creep in a swamp-water haze. Henri Matisse humored Picasso, the worm, as necessary and only just enough to draw him into false confidence while hiding the fact that the flamboyant Spaniard mattered not at all. Not to Matisse at least. Behind Pablo's back, to everyone else Picasso - the bludgeon, the conniver, the old woman - was at worst dangerous even deadly. The occupying Germans of WWII recognized this about the egomaniacal painter even before making of Paris a Guantanamo of the arts. Their deal with the Spaniard was a straightforward, clear contract of collusion. Recognizing their equal in demonic choreography, the Gestapo actually came to fear the manipulative sneaky little creep. Fear Picasso, but use Picasso. Before all of that, back in the shaky beginnings, it had started with how easily the bastard of Barcelona had rid himself of his competition for adoration, the new fragile angelicus of painting, Jules Pascin.

All of these discovered bits of history's interior fueled Jim Parks' push to find incontrovertible proof of the crime. The energy expanded to any other back-story of art history, expanded to make of him an astute observer of the sewer that is today's art market world.

Somewhere in these detective wanderings a few things clicked together and Parks figured out what needed to be done. He created a website as a virtual art gallery, not for commerce but for the alternative presentation of the works of artists, alive and dead, along with a high speed blog of art criticism most often aimed at critics, gallery owners, curators, museums, auction-houses, art historians, and art manipulators. The objective? Simple, wreck havoc where he felt havoc needed wrecking. Point flashlights into the seedy closets of those who arrogantly pursue control of the art world. Why? Because the evidence was clear to him that the art world had gone the contrived and massaged way of investment banking. Why? Because above all else, above war mongering, above environmental banking, above the New York Times' self congratulatory culture gang-bangs, above the rush to control food, above mining interests, above high tech wrestling, above all other quick manifestations of absolute wealth there rose art and art collection. And without a flushing of the art world's large intestine it was sure to die soon in a swill of its own bile and excrement.

He would become the Louella Parsons of today's art. Agree or disagree, the sorry art world would come to him and his website to get the latest dirt. He, his alias, wouldn't actually be the great equalizer but he would set the stage. The equalizers would be the Japanese, the Arabs and the Chinese when they found out how they had been duped over and over again by the art hierarchy. The equalizer would be the artists themselves when they learned of the arbitrary and obsequious way they were viewed and treated by the marketeers.

Social networking, that great small-ego fermentation experiment was being back-slapped into submission by Parks' maneuvers. In a matter of a few months the art world was a-buzz with the new "Commandant Marcos" of art and how he managed to unearth *this* stuff. The answer was all too obvious. Parks took notes of what he observed, fleshed out what that *might* mean and added in the heat of the most bizarre conjecture he could imagine. Only once had he been proved wrong, once. Watching the rats scurry across the deck of the art market whenever his reports came out told him he had found both the curve and nerve. Parks was not a good man, neither was he entirely evil. He was a nasty man who took no prisoners, who suffered no snobs or charlatans. Parks was a deliberate man impatient with consequence, his own or that of those others.

Anomaly we tag as a mess? China is certainly not a democracy nor is it purely communist. Autocratic China is developing a whole new complex system of governance which we might label as a 'contentorist' government with leaning toward occasional arbitrary balance, dependent as it must be on the gravitational pull of a demanding, openly-bribed politik and small-minded bureaucrats.

Gophers. Farmers and ranchers hate gophers. No news in that. Benny loved gophers, they were money in the bank for he and his brother, Swirl. Benny Hubris marked the day, in 1985, when he first heard that gophers loved chewing gum, especially that which had already been chewed. They loved it and it was deadly to them. They couldn't digest it, it plugged them up and killed them. That information stuck in his brain. He thought about it when he mowed his hay, trying to figure out how he could get lots of gum bits to put into gopher mounds. It was his brother Swirl who reminded him that theater seats have lots of gum stuck to their bottoms. It took a couple of years for Benny to figure out that his disoriented old cowboy-hippy friend, Stu Jeff, was the answer. In a raspy loud whisper of a voice Stu observed, "Gol' Ben, you'd make a fortune ifn' you could invent yersef a battree-oporated putty knife which a guy could use to scrape bird poop offn' truck hoods." Battery operated putty knife? That struck him as funny... then for some reason he had a vision of acres of gum stuck under hundreds of theater seats all over Oregon and Stu Jeff crawling along with his battery operated putty knife popping it all into a waiting zip-lock sack.

It was Swirl who suggested they market this stuff as Nate's Gopher Bait. It was Benny's light bulb moment to blog about it on harmoneyhunter.com the social networking website for Neo-epicurean apologists. He figured correctly that the more confused and convoluted the community the more likely they'd buy into the fun-side of imagining little gophers choking on gum deep in yard and pasture holes. The motto of the neo-epicureans was "If'n I cain't feel your pain I cain't not do it," and, "I never met an intel-

lectual who tweren't a fellow apologist." (The movement started in the deep
south when a college newspaper staff stumbled onto translations of nasty
renaissance writings when researching Machiavelli's "Anarchy for Fun and
Profit.")

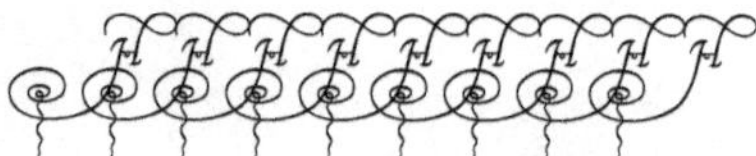

   Homo proponit, sed Deus disponit as Thomas à Kempis wrote five centuries ago. *Whatever you may dream, fate has other plans in store.*

   Up near Moose Mountain, not far from Sweet Home, Oregon, there
exists a wilderness area so well hidden by geographic elbow, recess, crag,
hollow, fog shroud and deadly precipice that few know of its existence. It
is called the *Menagerie Wilderness* because it is home to spectacular spired
geological formations like Rooster Rock, Rabbit Ears and the Turkey Monster. It's not on the way to anywhere; sheltered, even sequestered, as though
waiting for another time when such things will have new magic duty. A
three-tiered rain forest carpets steep undulations and hides the bases of rock
formations that shoot skyward as though signal perches to ancient flying
dinosaurs and none so wonderfully grotesque and unlikely as Turkey Monster, a three hundred and fifty foot tall pinnacle whose column tapers to the
bottom suggesting eerie instability. Three hundred feet up, in a fissure at
the widest part of the column is a significant cave which is summer home
to a monkey-man named Ashwan Clevalure.
   After his wife disappeared in this forest ten years before, little agile Ash
taught himself rock climbing to scale the columns of Menagerie looking for
a vantage point to find her. What he found instead was a breeze-laced calm
painted in washes of forest odor which gave to him her certainty. Sitting,
gazing from the top reaches of Turkey Monster he never failed to feel her.
   Hauling pieces at a time, he had outfitted that cave as a spare survival
chamber in which he spent four to five months a year. No contact with

the outside world. To improve his access, he connected lengths of khaki-colored, knotted parachute rope and tucked them, out of sight, in the spire's vertical cracks. Fastening baskets to the bottom of the rope, which hung twenty five feet above the ground, he eventually hauled up supplies by retrieving the rope from the top. Whenever he went down, he used a second rope fastened by slip knot release to the first rope, so that, once down, he simply jerked the knotted climb rope to release it to fall free.

Each Fall, with the first frosts, he would descend Turkey Monster and travel into town to rejoin his efforts to inject some sanity into a feckless world. He would suit up and punch in at the law firm of Comstock, Bench-ley, Clip and Schooner taking six months worth of only those cases where corporate malfeasance threatened to bury the disenfranchised. Ashwan Clevalure, legal champion of the small and frightened, monkey man to an abiding memory of love.

In that dream
left my hat behind,
moss came
from the sound of water,
thorns and brittle sticks
in angry wool,
a silent percussive
trail of rhythm.
There wanted to be
squeals for every
pin prick
yells for jabs in the eye,
but the wool muted
and no one thought
to ask for less

They met in the pumphouse on Jackson's ranch. Three ways to get there, all of them hidden from view in crack, canyon and thicket. A place to park was a quarter mile away where the sandy shoulder of the road was pockmarked with deer and human tracks funneling toward the low, loose, wire fence and the stile. That was not the way to the meeting place, that was the way east and opposite towards the grassy canyons where people went deerhorn hunting and poaching. On the west side of the road was a floor of lichen-covered scab rock leading to the top side of a short run of rim rock cliff. Enno knew the way, leaving no tracks, it was possible to pass round the junipers and down that cliff to a stretch of shelf midway below. Walking the shelf led to a long thin crack in the rock joining several shallow lava tubes. It was ten meandering feet deep and might take a determined, thin person to a gravel slide which accessed the ranch and the willow thicket. That was where the pumphouse stood.

Sloop watched from his vantage point and saw the young man arrive at the shed. He waited fifteen minutes and then quietly walked a quarter way around the perimeter. When he saw the doe meander across with no concern, he knew no one else was near. He joined Duden.

"Have you got any place on the other side where you might go for a few months?"

No hello's, how are you's, just straight to the question. They were nervous with each other, nervous to be standing this close again.

On the roof of the shed, the colorless Titus Ibid smiled to see them in the same frame. His foggy ghost innards were anxious.

Duden stood looking down at his feet, thinking of Titus.

"I know why you are asking. It's because of this new shooting. I'm not going anywhere."

"They want you and me dead. They can't get to me unless it is through you. Others will be hurt. It will be a repeat of last time. If you won't go then Helga and I will go - see if we can draw them away."

"You shouldn't have to go. I will find a place. I will go."

"When?"

"Now. Today." And he looked up, felt the strings of pain and walked out of the shed and back through the crack in the ground. No one saw him save for Sloop whose calloused heart was nevertheless broken.

Cruelty is the first and most obvious trait of the highly efficient.

How the Robin Hood principle seems to attract competition within the ranks, competition and the excess that comes of a misuse of imagination.

Within the hearts of a curious few honorable cowards, resides the ridiculous and worthy impulse to crack heads and spread wealth anonymously.

Penny Frill froze. There it was, the clear view. Long ago she had realized, when wearing sunglasses, that fragments of images were always moving past her at the furthest reaches of her periphery. If she looked sideways, whatever the image was disappeared. If she looked straight ahead and pretended not to notice, allowing herself the widest wide open visage, then on the edge where her vision almost quit, there was all the moving, telling imagery of some parallel universe, some flickering, completely functioning ghost world. She could see just enough of it to know it was there. Not enough to get an accurate picture. To anyone else it would have been a reason to make a doctor's appointment. For Penny it was a thrilling reassurance.

Only this time it was different, more, and clear. She could see the black and white image of a skinny old man with a wispy white beard. He was laying across the top of some sort of metal trailer and beside him lay a full color, ragtag, old cougar.

She said to herself, "I can see you." And she was not surprised to hear, in her head a response. "Yes, and I can see you."

That was how Penny Frill and Titus Ibid began their first conversation.

Jeb, the hit man, drove the freshly rented Subaru wagon down the same forest service road again at 40 mph, getting it properly dusted before returning to Mascara's downtown, only to find Fred English looking at him and shaking his head as if to say "this don't work either." Jeb said, "Shit" again.

Chapter Seven

the dry vase

*The enormous encyclopedic work in China of
the Four Great Books of Song, compiled by the 11th century
during the early Song Dynasty (960–1279),
it was the massive literary undertaking of the time.*

*The last encyclopedia of the four,
the Prime Tortoise of the Record Bureau,
amounted to 9.4 million Chinese characters in 1,000 written volumes.*

*"We had the liquor, the chicken, the music and each other, and had no
need to pretend to be what we were not. This is the freedom that one
hears in some gospel songs, for example, and in jazz. In all jazz, and
especially the blues, there is something tart and ironic, authoritative
and double-edged."*   *- James Baldwin,
Letter From a Region in My Mind*

When a confused and angry Enno got back to Riven's cabin, Benji was gone. It added to his bad mood. He was pleased to see Resumé there, standing guard on the porch. Once again his dog was chewing gum. There was a piece of paper pinned to the door. In an elegant hand it said "come to lumberyard, bring rope and carpenter tools." It was signed Aboo. Here was something Enno could understand, something direct and purposeful, even if he didn't have a clue where it was leading.

Promto Manks, alias Jim Parks, got the message from Sloop. Time to tool-up the hacking armory. His cohorts called him PM for short mostly 'cuz he only worked after dark or leastways they only saw him after dark. And what they saw didn't reveal much. They could tell you he was thin, average height, red hair and spoke with a lisp. Plus, he always wore a thin black zorro-style eye mask. What they couldn't know was that the eye mask was painted on, the hair was a wig, and the lisp came because after dark he took out his upper denture. PM did his hacking work out of his Dodge service van in which he had three electric vari-desks with three custom-built steam-punkish, mega-powered computers each with three clusters of three wood-framed LCD screens and bracketed by an analog component stereo system with turntable and two large Infinity Studio-Monitor speakers. The outside of the van was covered in software code and chinese characters. (In another life Promto had been Chinese crossed with Egyptian. In the daytime, and as Jim Parks, he was the maskless and balding art critic with a blond monk's fringe. Denture in, he spoke clearly but with a Swedish accent he absorbed from his step-mother and her boyfriend, Victor Borgé.)

Promto drove his van, at sunset, to the outer edge of the Walmart parking lot in Bend and set up his folding chair and hibachi grill out front, along with a *Bill Murray for President* banner, hung off the unfurled side awning. In the hibachi he put a small bag of puppy poop and lit it on fire. It always worked, everyone stayed as far away from his van as their noses found appropriate. PM put a little Vicks Vaporub in his own nostrils and climbed in the van to work on his computers. First order of business was research. 'So, who now is sending hitmen to Mascara?'

Planet earth is a living thing. And planet earth is ill. Its biodiversity is being destroyed from within. All of Earth's life depends upon this biodiversity. Human kind is destroying the living skin of this fragile and precious planet. And human kind does not care. L. J. Shoulders, the old sculptor and rancher wants out of the human race. His is a crisis of purpose. How does he justify making sculpture when he feels to his bones that HIS planet may not outlive him. He hears the legitimate concerns of so many of his poor and hungry fellow humans and, on the one hand, feels their suffering while other often times hearing himself deride their wounded self importance. Humans are instrumental... No, no, mustn't think like this...

At the lumberyard Benji stood in the parking lot with three sheets of paper, itemized receipts. He beckoned Duden to drive round back to the stacks of boards. He explained to Enno that they were going to pick up lumber and supplies to begin, together if Duden agreed, to build a shed on his Snake Flats acreage.

"I can't afford to do this now," the boy said.

"It's already paid for," answered the brown dwarf. "While you were gone I saw your sketches for the cabin on the table. These boards will give you a start on that building. I want to help, I talked it over with your dog and with Titus."

"No, no, Titus is dead. And I promised Sloop I'd go away for several months."

Benji looked at Enno with head tilted and slowly said, "And so you shall go away... to your land and your own cabin."

"But..."

"First we need thirty-five of these 2 x 6's. And then forty of these 2 x 4's...."

As they loaded lumber, Benji listened to Duden's worried slow tale of invading assassins and Sloop's argument for Enno leaving. Benji asked him to repeat the earlier story about Dr. Fil and the well or mine shaft. He would need to spend time with the doctor. Resumé sat in the truck cab and methodically chewed his now flavorless gum.

(So long as markets be damned and books be allowed to live unread, as the narrator I am to have the liberty of inserting asides which confound and perhaps illuminate all at the once. Such as:

"...the ruined castle of La Mothe-Chandeniers, in Les Trois-Moutiers, central western France. Specialized in the rescue of old stones, the crowdfunding site dartagnans.fr and the association Adopte un chateau launched at the end of October the collective takeover, via internet and by mutual agreement, of the ruined castle of La Mothe-Chandeniers, for 500,000 euros. As a result, some 6,500 internet users responded to the initiative, and are now the castle's owners. GUILLAUME SOUVANT / AFP."

What better aside than the image of 6,500 people throwing 500,000 euro coins into a fishbowl marked 'Castle the Storm!')

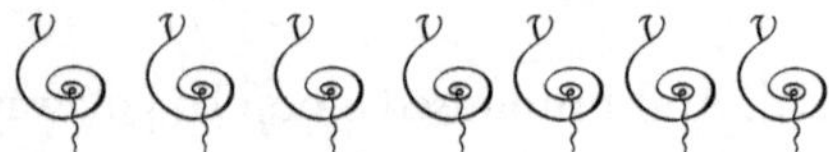

Way back on a previous page, Bessie Quince was making her initial approach to Houston Tuttle, the motivational speaker and lust detonator. To watch her float in, an artificial sponge-bird claiming the desire out of any room, you would never have guessed her weakness.

Bessie Quince, the svelte, made-up beauty and New York arts community social climber was once a stuttering child, not in the oral sense but in the physical sense. Long ago, circumstance used to turn her normally, naturally, lovely self into an ugly, wall-eyed, bumbling creature. When nervous or excited various parts of her body would twitch violently, might be a hand or leg or cheek muscle any of it causing her to lose coordination. She would, of a sudden, have trouble walking, stumbling side to side. When she had it real bad her forehead would twitch and her left eye would wiggle independently and point itself out in an effort to deny partnership with the right one. Strabismus or Tropia might have been a diagnosis, but her mother, a yoga instructor discovered that exercises and determined mental focus seemed to cure Bessie. She hadn't had one of those episodes in fifteen years. The memory of them, however, sent cold chills up her lovely spine.

Right now, as she floated towards Tuttle, people stepped aside and prepared for what would certainly be the spectacle of Bessie 'doing' Houston. He was shaking an older woman's hand and looked up into Quince's perfect brown eyes when it happened. Bessie Quince saw gorgeous Houston up close and her eyes went wide and separate. Of a sudden she became one ball of spastic twitch, her knees buckled and she stumbled, lurching forward. Tuttle caught her, and as she rolled in his arms he looked down at two twitching wall-eyes above a drool-laced, open mouth.

Bouncing back up on her feet, she waved her hands frantically around her face as one would brush away a swarm of flies. It had the compelling effect of insisting to all who were near that her twitching, wrinkled, squinty impatience was evidence of some sort of terrible contagion. And Houston found he couldn't look at her for long without being compelled to twitch and squint in response. Everybody backed away.

When she realized Houston was a mirror of how she felt and looked it made her angry that he would dare mock her. But when it was obvious he too was out of control, her face and countenance relaxed and she burst out laughing. The same happened to him. In one last vestige of the nervousness, she stumbled again into his arms, and this time it was all over for both of them. All who surrounded the couple hugged themselves.

(And they say you can't make this stuff up. Of course you can. But first you/I, the narrator without portfolio, must blank out of your/my mind any notion that serious readers ever meet here. This narrative does not pretend to be serious or incremental, it is delirious and excremental. This is not the substantive confession, that comes later, much later.)

At Enno's acreage the men sat on a rotting log thinking out loud in two sometimes parallel, sometimes intertwining, lines.

Duden thought: *This creek is new. The ground is being carved and pushed around by this fresh course of water even as we sit here. I'm going to follow it upstream and see if someone has left a big hose running...*

And Aboo thought: *Just as with this young man, this perfect agamogenesis, this apomictic, parthenogenetic, agamic salvation, here are waters so new, waters without history, that they are shaping the future right before our eyes.*

Enno looked in all directions. "Before we do anything else, I'd like to walk up that way." And he, and Resumé the gum-chewing dog, stood and moved along the meandering waters.

Benji watched and, instead of following, went into the old, stone pump-house remains to check out the big well hole. Dr. Fil's block and tackle and ropes were still there. He squatted, closed his eyes and templed his fingers. He sat that way quite a while until he felt Resumé nudge him and knew Enno was standing nearby.

"We found..."

"Yes, I know. You found where the water comes bubbling out of the ground, surrounded by dry sands and rock. You found affirmation."

Shaking his head he said, "I don't understand, this is wrong..."

This time it was Benji's turn to wave off the discussion, "For now, what about we unload this lumber here by the well site. I've got an idea to share with you. I think we need to take some action to make sure this new stream does not flow down into the well hole."

"No, I want to unload the lumber up near where I'm going to build a pond. But yes, about the water, I've been thinking about it also."

From his perch, beside the reclining cougar, in the big limb-cradle of the Ponderosa pine, twenty feet above ground, Titus looked down and smiled in stanzas with a curvilinear, narrative chorus. He felt his ghost-self in perfect communion with the lingering afterlife of Hoagy Carmichael. The old, slowly dying cougar/mountain lion/puma/panther/catamount/felis concolor watched the two men unload the lumber and purred silently. He could feel Titus stroking his fur and was untroubled by not being able to see him. The brown dwarf had joined in telepathically, reassuring both cat and ghost.

Benji nodded at Enno's selection. It was a most suitable place to build a cabin and a pond. Others may have thought so once. Off to one side the perimeter was lined by the crumbling remains of an old rock wall. And an embankment seemed to frame the memory of a snow-melt catch-basin. Also, there were close-in, intimate variations in the landscape, as though a miniature mirror of all the wider, forest landscape which surrounded. Cracks, swales, rises, hollows all on a half scale as if God had pinched this corner of the earth's covering.

"Your excellent land is home to wild bunch grasses of great good health. I suggest to you that this may be a clue to how you can disappear right here but of course not without help. Perhaps it might involve allowing my uncle Nimbo to join us for the season with his tailed Jacob sheep flock and his guardians. We also may need access to a backhoe."

Benji explained how it was that Nimbo, a retired desert sheepherder maintained a small flock of thirty-seven ewes and two rams and traveled around to rented and borrowed pastures. With this band of sheep he kept three rare guardian dogs of the Transmontanos breed. A large, fierce, breed developed to take down wolves and coyotes. Uncle Nimbo had raised and trained these dogs to keep any designated thing away from the flock's domain, man or beast. The trickiest part of bringing the Nimbo family to the acreage was going to be Resumé. Benji had to be sure the Cao de Gados wouldn't kill Enno's Aussie. Wise as Benji was, he had no way to know that Resumé's mastery of bubble gum would be the equalizer.

"With Uncle Nimbo we accomplish two important things. Number one, NO ONE will be permitted onto the land without Nimbo's release, the dogs will see to that. And number two, the sheep will gently discover and prepare the ribbons of best soil on your new farm."

Long pause, "I think I need to be alone for awhile."

Benji rode off on his motorcycle.

It took some doing but Benji learned that Dr. Fil was at the Mascara library using the public computer to access his emails and some scientific files. Penny Frill was at her librarian's desk communing with a new ghost she found more interesting than most. Benji looked at Penny, Penny looked at Benji, and Titus whispered to her, 'that's him." Penny thought, *of course it is*.

Benji smiled 'cuz he heard Titus's words as he felt Penny's curiosity.

Titus could read Penny's mind. Penny knew it. Titus could not read Benji's mind. Benji Aboo couldn't read minds, but he could visit most on command. These were sloppy, overlapping circles which would need to be controlled somehow.

Benji waited, hands clasped before him, for an appropriate moment to speak to Dr. Fil, who looked intently at the computer screen, pursed his lips, and raised his eyes, slowly turning his head.

"Yes?"

"Excuse me, sir. Would you be Dr. Filipiano?"

"Yes."

"My name is Benji Aboo. May I speak with you privately, sir?"

The scientist smiled at the pleasant formality and extended his hand to the little man. "A pleasure to meet you, I am sure."

They stepped outside and sat on a stone wall around the library flower garden. Dr. Fil wasn't tall, he was a leaning-long in that way of Meerkats, prairie dogs and Indian Runner ducks. Black-skinned Benji on the other hand was short, simply and perfectly short. Only his spectacular nose was out of proportion but certainly with purpose.

"I am a friend of the boy, Enno. I have just returned from a visit to the hole in the ground."

Fil straighten his back and gained feet in height. He looked straight out and away, well over Benji's head. His brain took a breath. Then instead of looking down to meet the dwarf's eyes he leaned forward, curved his back and lowered his own eyes until, without threat, they were straight across from Aboo.

"Yes?" asked Filipiano.

"I want to understand your equation." Benji hoped he could smell in this man a motive he might trust. Hoped. He seldom found security and certainty in hope, seeing as they were divisible only by emotion.

The narrator all a jumble, losing his place. Clothing suddenly too tight.

Duplicities - outsider devotion to private working - clumsy efforts to social-
ize - feels himself morph to poetic truth without guile, without prejudice,
so long as his space is vacant. Woman needing also to belong to herself.
Belonging completely to self while devoid of any self-centering - no spite -
no others to impose compromise - no excuses save those which allow them
to continue.

He thinks, forgetting his job, humans are as various in personality as are
insects of the rainforest and as are the fringed fluctuations of the universe.
Discover pattern, a comfort, a predictability and stress it to the nines - or
don't, just give it time. The surprises, disappointments or thrills are ours.
Nature? She does not see them.

Then he pretends he doesn't care that he has dropped the ball, and returns
to the story.

Back eleven months ago, state senator David Actualroger had taken the
afternoon off for the first time in six years. Here he sat at the college coffee house in Corvallis sipping great java and picking at an atrocious, dry,
surprise-less pastry.

And all the news kept piling on his head. One tragedy after another,
deaths piling high, the White House making inedible sausage, vicious class
and race hatreds run amok, diplomatic intrigues in idiot garb, fevered and
fragile markets, greed and avarice a blanket for as far as anyone could see
and feel. His sense of purpose ached with destructive pressure. Something
had to change. He thought: *It's not what I signed up for, all those years ago in
college. Look what I've become. A cog. Governments, even little state and local
governments, are the single most destructive force in the world today for they
legitimize, rationalize and even deify profit-taking and, in so doing, speed the
devastation of this planet. And, damn it, they also give permission to lunacy and
control. Morons top to bottom. And I'm one of them. Governments have become
a bastion of small and mean-spirited minds. Ours, state and federal, are the
worst by 'titanic' measure. Where's the imagination, the heart? How did it come
to this? I think that man was right. Governments have gotten too big.'*

He opened his eyes.

In that sober, half-baked, crumbling-pastry moment senator David Actualroger came up with the idea for a state referendum. It would call for Oregon to split from the United States of America. And he would not start with his colleagues. No, not with the dunderheads. Instead he would start with eighteen and nineteen year old students looking to make a difference. He would start with social media. It was to be a campaign for true change and it had to start with Oregon becoming its own separate country. That was eleven months ago and today the murmurs had taken multiplying form.

And then his colleague, Jim Hardcliffe, sat down with his cup of synthetic Jasmine Elderberry tea made from frothed and thrice-purified recycled urban water.

"I see where your campaign to secede from the union is gaining speed. You should be ashamed of yourself, getting all those unwashed children out on the streets marching for something so destructive."

"How do you figure it's destructive Jim?" He regretted the question the minute he said it.

"Listen Davey, the way the world is going now everything is out of balance. For example, did you know that there is just one country on this planet where the population is declining? Guess which one, guess. Okay I'll tell you. It's Russia, and you know why? It's because every attractive Russian woman of breeding age is being bought up by we politicians and those industrial chiefs and brought to this country. They're here illegally, Dave, and those women are using our services to have their half-breed babies. Pretty soon the only people left in Russia will be those old Russian mobsters and broken down ballerinas."

Senator David Actualroger stood up and walked off, shaking his head the whole way.

Meanwhile, Jeb, the hit man, drove the freshly rented Chrysler minivan down the same forest service road again at 40 mph, getting it also properly dusted before returning to Mascara's downtown, only to be pulled over by the sheriff who wanted to know how it was he, Jeb, a stranger, had been spotted driving three different rigs downtown, all in 24 hours. Jeb made up a good story about mechanical mishaps, said 'Shit' to himself, and drove on. He was damned lucky that chrome case had been under his coat in the back seat.

Enno Albert Duden and Resumé meandered across their new lands, both looking with relaxing eyes at random spots and vistas. The young man invited his brain to wonder after what he saw.

And the old dog pushed his piece of ABC gum up to the front of his mouth, where his front teeth teased the elastic lump until it stuck to the gaps in those teeth. Slowly pulling his mouth open, he created a thinning wall of the gum into which he breathed normally. The gum wall expanded out and then gently collapsed back with each breath. It was a bellows or diaghram action and it made the old dog very happy. He was understanding new dimensions of cause and effect.

They came upon a shortened, emerged length of rim rock, a wall ten feet tall and sixty feet long of red-grey and black rock stacked as though by toss, not from men but from some long ago subterranean push. There were cracks and small shallow caves punctuated by wild bitterbrush and willow. Enno noticed a curving pine growing as if out of the top, rising from a crack in the rock. Down below, there were several rocks that had apparently fallen from the rimrock. He went to one and sat down, back to the wall, look-ing out across the landscape. For as far as he could see there was no recent evidence of man, just the characteristic scatter of a desert forest and wild grasses. He felt genuine comfort.

Resumé smelled something and went to the rim rock. Below the spot where the curving pine rose, and behind some shrubs, there was a tall sepa-ration in the rock that opened sideways. With hesitation the dog went in.

Enno heard Resumé's yelp and called to him. No answer, so he went in search. Yelp again. Pushing willow aside, he was able to stand up and walk into the dark crack as he asked for his dog to respond. Now he heard whimpering. Hands against the rock surfaces, he continued until he saw a faint light back deep in. Something told him he shouldn't be in here, and he wouldn't have save for his dog. He had no flashlight with him. His eyes were adjusting to the faintly lit dark. There was a ten foot wide bulging rockface, like the belly of a giant toad, which curved and tucked to create a four foot deep cove at the floor. Enno made out Resumé's butt and bent down to see what the attraction was. He couldn't see into the depths of the cove but soon all was answered, as out poked the snout of a tiny coyote pup, growling and snapping at both of them.

"Hey, buddy, we need to leave here before mama comes home." He patted his dog's back. Reaching up the rock face for support he felt parallel grooves in the rock. Light so faint, he could barely make it out as he rose, but there were etchings and painted figures on the rock. They looked old. He made a mental note to keep this discovery to himself and to come back here with a flashlight. They were making their way out of the cave when Resumé stopped, hair raising on his back. Duden had the presence of mind to keep quiet. Up ahead some thirty feet, they could make out the head and shoulders of a cougar sniffing the air around the cave opening. Then they immediately heard what sounded like many coyotes yapping feverishly and they saw the cougar turn slowly away from the entrance. The cougar screamed and jumped. It sounded as though the coyotes were attacking it. Duden and his dog used the diversion to leave the cave and move off away, as quietly as they could. Invisible, Titus watched them pass by.

Smiling to himself, Duden nodded that this was exactly where he wanted to build his farm, his home, and his life. He didn't understand completely, but he knew it to be so.

Dr. Bone Stock, with his luggage and his two styrofoam coolers, rode in the sleeper cab of the D.O.E. oil pipeline test vehicle. They were waved on through, unchecked, at the U.S. customs on the Canadian border with North Dakota. He was anxious to get back to Oregon and the Warm Springs reservation where he had two expectant mothers waiting.

Dr. Bone Stock is a native of Moosejaw, Saskatchewan, and credits his humanitarian interest and love of intrigue to the fanciful history of his home town.

"A man should keep his little brain attic stocked with all the furniture he is likely to use, and the rest he can put away in the lumber room of his library, where he can get it if he wants it." - Arthur Conan Doyle

"He imagines where nothing is false, and he isn't tossed about as things shift back and forth. He knows that the endless transformation of things unfurls according to its own inevitable nature, and he holds fast to the ancestral source."

- Zhuangzi from the Zhuangzi 400 BC

Deep secrets
not of deceit or theft or failure
but of beauty's equations
oily grasses of mossed steam sawmill
stove wood stackings perfection
oxen way of working horses.
protected sheep on sun warmed rock
young elk walking on back legs
old ignitions begging for the tuneup of forgiveness.
mushrooms 'neath the rotting needles
old books in need of eyes.

In the dream there was a barnacle-encrusted hovering leviathan filled with self-generating helium-like hot gas. It was a living thing in as much as life lives. It had the shape of an enormous Russet Potato with a similar surface. The low areas were wet, the upper reaches polished, betwixt were crusts. Through a microscope you could see mites scurrying to work down backlit highways that led to cities where better dressed mites sat on courts which sent mites of unfavored colors to prisons within the crusty parts.

Part Two

jaquima

*"We have machines to scan the flesh and
track the blood, game the stock market, manu-
facture our news and social media, tell us where
to go, what to do, how to point a cruise missile
or a toe shoe. Machines neither know nor care
to know what or where is the human race, why
or if it is something to be deleted, sodomized, or
saved."*
- Lewis Lapham 2017

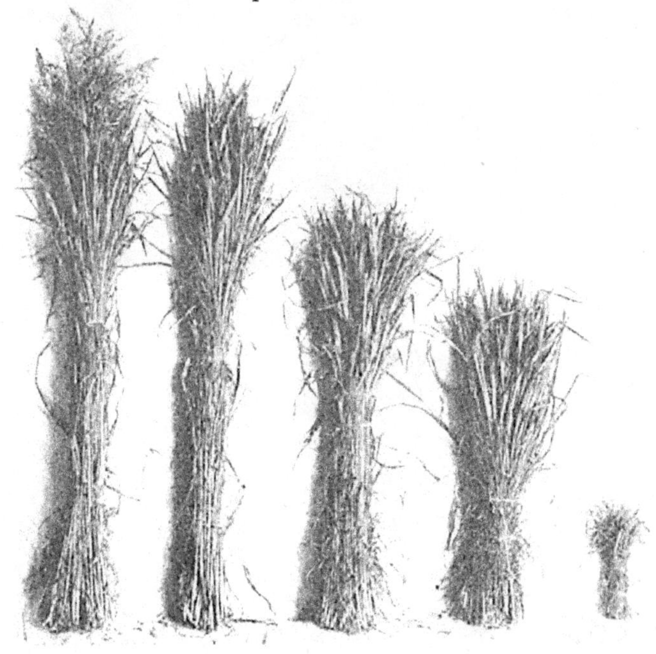

vibrations hidden
only through resolved resonance
state the music,
mulattoed pinched nerve
of art's answer

Chapter Eight

k n a p w e e d s

Schadenfreude: that malicious joy at
another's suffering or misfortune

the acceptable long sentence, the Proustian gamble
the head-aching vibrating mirage
- J.D.Stubblegripper

Benji found Enno back at Riven's cabin, working on sketches of a building. He set a bag of groceries on the table and gave Resumé a large dog biscuit. He unpacked a bag of dry adzuki beans and another of mung beans. Then there were bunches of herbs, a cucumber, brazil nuts and some honey tangerines. He peeled one and set it on the table by Enno.

The young man watched, then said thanks and went back to his sketching. They sat together quietly, Benji paying close attention to the growing cabin sketch. Enno tapped the pencil point on the sketch and got up nodding his head. Together they went outside and loaded the carpentry tools in the back of Duden's '52 Chevy pickup for the 20 mile round-about trek to Snake Flats.

Enno had spent two years working for a general contractor in Eugene and had absorbed information for just such a day. As he put in stakes for corner string boards, to square up the site for his cabin, he was surprised to find that Benji followed and joined with every step, proving he was fully aware of the process and intent. The short man stretched out the 100 foot tape so they could measure diagonally from corner to corner. Matching numbers would square up the footprint of the site.

Enno held a long level along the string to determine the lay of the site. With no words, Benji came over and took a small string level from his shirt pocket and hung it, middle of the line-course, and pointed for young Duden to understand. Then Benji nailed three long sticks into a tripod and balanced the long level on it. He sighted down the level. Enno instantly knew what it was, a make-shift transit. Benji Aboo walked to the other side of the site and hung a tape measure upside down while he pointed to the numbers with his finger. Enno sighted down the level and pointed up or down until they found the right height. With this process they determined that the selected building location was level within three inches.

Taking a break, Benji once again brought up the subject of Uncle Nimbo coming with his sheep and dogs. The idea was for Nimbo to graze while helping to protect Enno and his land. He also told Duden of his meeting with Dr. Fil and the idea of expanding their little circle of protection and confidence. It was more question than suggestion.

Resumé sat in the shade, nose up and eyes closed as he kept an olfactory 'watch.' He wished he had a new piece of bubble gum.

The odorless ghost of Titus, hanging midair, was enjoying the view of the work when he received a jolt and knew he had to leave immediately.

'This narrator is a silly idiot', writes the author.
<mirror>
'And you think you can do better I suppose?' Asks the narrator. 'You know, Snidely, it is seldom a good idea to have the author show him or her self, right out in public. That's like representing yourself in court, a fool's errand. And to deny that a fully invested narrator can exist without imposing his own character on the pile of words IS silly. To worry about what casual readers might think is to deny four-tenths of the adventure. In fact it may deny the full story ever gets told. But go ahead, Mr. Careful Know-It-All, you take over.'

<mirror>

The author writes, the narrator narrates. The author creates or finds a 'suitable' narrator. Or the author imbues a character, thought to be fully understood, with the power to tweak the story and allows that character to prance and mumble, gesture and fumble, at stage left with a pen light and a slimy glass of Idaho Vodka.

<mirror>

'You think because you write it you make it so? Idiot. You have this long story in your head and you micro-manage it towards your revelatory goal or climax. I don't have that story in my head. In fact I don't have a clue where this is going. I am free to see and record and announce only that which is right in front of me. That makes of me more a part of this story than you will ever be. Writers are so over-rated, so pompous and self congratulatory. Just because you give life? Give me a break! Writers at their very best are shepherds of words, working to keep them together and headed towards the gate. Readers don't care about Melville, only Ahab...'

<mirror>

'Okay, I'm done with you. You are dead to me.' The author spits these words and goes to the toilet.

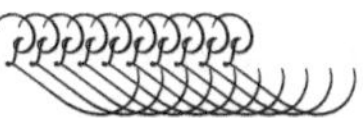

Some lives are aimless and avoid all forms, complexities, responsibilities. Others force form on themselves to odd ruination. The fortunate ones find their ship of form, of discipline, of blended belonging mystery.

Gary Snyder: "Wilderness is a process."

Jim Harrison: "We live in particular."

John B. 'Sloop' Ogdensburg: "My own character has gone to insular cruelty and selective insensitivities. I shunned even myself."

Some older people (or not) lose their resilience and become unwilling to consider any major change. Does that come from not being able to see a future with which to work it all forward? There is the finality of this life - but young people with less of a clock-tick also appear to have lost resilience. We permit the clock to rule us. How to deny permission and see it all as bonus time? How to clean everything up? Planting, always planting. Building, always building. Creating always. Working solidly towards a handmade life wherein the grid and the tyranny of complete support systems fall away. Where might the new community reside, where does family extend to?

The old man wore a black gabardine, snap-button, long-sleeved shirt with oily food stains across his chest and stomach bench. Posturing to be the leader of the family group, he lead them to the queue. The doorman said 'no - go to the back' and glanced with rolling eyes at the stained shirt. The old man turned away so that the doorman would not see him seeing, so that invisibility would be given a chance. Frustrated, the doorman came up from behind and took the gabardine elbow guiding the old man to the end of the line, then turned him to face in the right direction. Looking down at himself the old man muttered too soft to hear, 'he doesn't understand, he can't know who I have been.'

The swollen teacher who hunts deep and cocooned talent to ride the metamorphosis, with the crazy notion that stages need staging, that colors need coloring, that arrivals need heralds shrouded and owed. What shelf the talent sat upon had more to do with its ability to scurry, backwards, up the wall like a frightened young cockroach save for the fact that cockroaches, everyone knows, are never young. Talent, forever prepared to run off.

He noticed in that instant the blood. He felt, for the half second, available to the pain. And the raw needful excitement of the job at hand returned him to fluid and to defining. Twenty-four was such a difficult age.

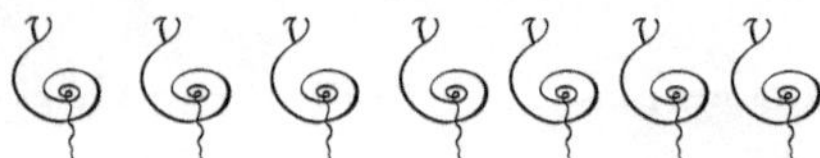

Promto steals a ride on the Oregon secession campaign and shapes it surreptitiously towards a nationstate which protects nature first and the dignity / safety of the human community a distant second. Instead he calls it the *detachment campaign* to form a more perfect nation: with a country-wide non-profit structure in which every business "leases" time and space from the nationstate: rather than taxes, Oregon would collect lease fees. And twenty percent of those fees would be earmarked for the preservation of biodiversity and natural balance. It's considered part and parcel of this nationstate's health plan. In Promto's nation any business or individual who fails to make its lease payment to the state forfeits citizenship and is summarily deported to the old frantic and dying USA.

P.M., a.k.a. Promto Manks, uses his *Tweaker-feed* like an exacto blade. With staccato bursts of 37 character suppositions, he lets young people know that there is no fixing the old system. It is time to build something entirely new. It was to begin with formation of a new political force, the **two-tailed-dog party** referred to as OOF (patterned after the Hungarian), promising everlasting life and free beer. A party which promised to dissolve just as soon as Oregon was successful in detaching from the union. He launches his "what if..." manifesto in bits and pieces with constant repetition until a quarter of the young people on the planet are frantic to move to the new all-natural nationstate of Oregon, where the totality of biological life is given free health care and all signage is done in fourteen languages, (oh, and by the way, advertising is outlawed). A nation where to be an immigrant is to be granted an all-expenses-paid scholarship to the community college of your choice and a propane-fueled hibachi.

Senator David Actualroger had no clue this parallel effort existed.

For a lifetime he carried around in his old head a mind - brisk, absorbent, curious, even humorous. A mind always far more capable than the host brain. He followed that mind. Now, each day he pursues that mind, knowing he must catch it unawares and pull one of its strings until it starts, once again, to roll. Then get out of the way and quickly record what it is

                    brown dwarf

thinking. Often now, that happens in the middle of the night or first thing in the morning.

Fred English, retired Oregon state cow cop, woke up one morning old. The very next morning he had received a message from a noteworthy nemesis, The Monk, and it got his juices going. He had someplace to roll to and for. That done, he decided no more waiting, he would go in pursuit of the puzzle's release.

"I do love a ripe Opal apple. But they are hard to find 'lessen you are in Opal apple country." Biting into the crisp, golden yellow fruit, Uncle Nimbo was talking to himself and his three big dogs, all of them crowded into the old Airstream caravan trailer, parked curbside, adjacent to the Safeway market in Madras, Oregon. The trailer was being towed by a '74 Chevrolet two ton 18 foot flatbed with stock-rack in which were Nimbo's entire flock of haired sheep. On the rack over the cab were six bales of 3rd cutting Alfalfa hay.

"Little Buddy says we might try the back way into Snake Flats. Slow go and rough, he says. He fergits those places we frequently go, huh boys?"

Uncle Nimbo's dogs were majestic *Cao de Gado Transmontanos* from Portugal, for all the world looking like short-haired, weight-lifting Saint Bernards. These canines had been developed over long Iberian winters to drive off, and even kill, wolves in the protection of their flock and homestead. Mikey, DeGaulle, and Sarkozi were only two-year-olds but their high level of training and willing obedience to Nimbo suggested they be much older.

And their keeper, Jim Nimbus Sparkle also known as Uncle Nimbo, was a walrus-shaped human, seven foot tall. His perfectly-rounded, constantly-smiling, fuzz-topped head sat on his broad shoulders with a telescoping neck. Turtle-like, he could extend his head upwards or pull it down to the shoulders as the mood and circumstance required. He was fifty inches around at the widest point and always wore a 4XL long-sleeved red-orange pearl-snapped chamois shirt that flowed, sail-like, down and into his baggy canvas duck trousers restricted rope-like between the tight verticality of his wide Prussian blue suspenders. Year-round he wore bungee-tied ice grippers on his size 23 Redwing work boots. No way was he taking any chances of falling down because it took a very long time for him to get up. Up he was fine, up he was great, up he moved as well as he wanted to. Up he was a powerful giant capable, with proper footing, to lift enormous weights. But down on his butt or belly he was worthless, nearly stranded. Inside his Airstream, over his bed, hung special stirrups. He could grip any of them and press a button that activated a little winch helping him up. He held on and reversed these winches to lower himself into his bed.

Giant Nimbo's smiling round head and enormous blue-striped red-orange bulk struck absolute glee in the hearts of small children, disoriented women and those afflicted with Down syndrome. His countenance never belied the fact that all of those needy people scared him poopless. He was happiest on the edge of remote forests, alone with his sheep and dogs. Very few people earned his trust. Benji Aboo was at the top of that short list because Benji understood completely. On this side was a seven foot tall, pink-skinned, strawberry blonde, fat giant of a man, and on that side was a big-nosed, four foot tall black Aborigine in a perfect upright frame and leather cap. When they were seen together the contrast was head shaking.

Nimbo never went anywhere, woods or town, without his Orion Safety Alerter Coastal Signalling flare gun deep down in his trouser pocket. The boys, his dogs, were trained to return to Nimbo's side whenever they saw and heard one of these flares go off. But they didn't barge to his side, they first circled a wide perimeter around their master, and the sheep flock, assessing threats, only then did they go in one at a time. This was their way of being best prepared to deal with hidden and impending threats.

"It's only life, Mikey. How difficult can it be?"

ʃՑՑՑՑՑՑՑՑՑՑՑ

Maude Smalldoll, the art critic for the New Yorker, was like an aging, punch-drunk, prize fighter in short-top, velcro-fastened ice skates wobbling to Henry Mancini organ music as he insulted any and all of beauty in a word-slurry race towards absolute insignificance. The man was so smart he had succeeded in emptying himself out. Inside of his brain was a spastic light show driven by a perpetual argument between Camus and Beckett about what to wear. The only amusement he got these days was when one of his sentences managed to bite itself in the butt. His one mortal fear was that no one would talk to him at the next squishy cocktail party. He needed people to disregard, he needed them to want to be disregarded by him, it was his poisonous fuel.

He was never the hero, always praying not to have to choose between his hatreds and his fears, understanding the best of him came when he failed his nature. All of his sacred promises were made for him, in his crying presence, by his...?

What do we feel, what do we allow ourselves to share to greatest sorrow? We slide down that other side to be caught against a tree of another's suffering. We slide because it is who we are, but it should not mean we forfeit the absolute right to pick our battles. At the core of so much is the nature of attention. Do they have our attention, do we theirs? Could they ever have our tears, would we ever want theirs? Politics and profit, both demon blood, and unspeakable as such. We must, each of us, never, ever 'get over ourselves' for in that transformation is membership in the church of excuse.

Pressured ballyhoo, the Oregon legislature considers a referendum to split from the U.S. They make the stupid assumption that putting it before the public for a vote would be one way to stop the silly discussion forever.

Ashwan travels: All across the mountains the vistas were exciting and invigorating, clear evidence of winter moisture's effect. Thick growth. The

Tombstone pass route is wondrously difficult as it hides the geological extremes of tight narrow canyons shrouded in moss and over-story canopy. Knowing it is there surely accounts for something, but the hidden Menagerie forest conceals all. Perhaps too much narrative to ever make of it subject for painting? Or? Afterall, the sea coast is example of too much and just enough.

§ § § § § § § § § § §

Amaroza, the Mojave's 'hide and seek' river, a complex subterranean puzzle punctuated by geothermal sources hydraulically ramming, pressurizing lava tubes. A clue for futures.

More people, but less people. The interactions seem gone. Conversations feel like nuisances. Such oddity. Two different mornings, a week apart, the lifetime waitress told me exact same story about the wild deer that visited her and about the Lupine. Same facial expressions and all. It was as if she long ago had given up on people telling stories to her. Doomed to sorrowfully repeat. Then the family restaurant sold and the spell was broken.

Wondered after how church, community and liberal arts education have been completely supplanted by the thin nastiness of 'social media' and that we are witnessing entire generations of rudderless humans. Don't think it is what 'we' bargained for. There is no value today in actual face to face conversation. Real time, same space, human interaction has always been divisible by zero. For this reason the metrics bust apart and monetization goes up in smoke.

<'Holy crapshoot. Without me you are little more than a *Don Quixote de la Pee Pee* wandering clueless and oily through the H^2Oed streets of this our new and wondrous digital *'univorse'*. I cannot, I will not, permit you to proceed without me, your filtering sidekick narrator, so let's make this deal; I promise to never roll my eyes or tonsils when you do your stuff, if you

promise that readers will understand when its me talking, rather than you. Deal?'>

Benji has a small chin, half-hidden behind a white bristle-brush moustache. Prominent cheek bones separate his eyes from his big ears, and they frame his large nose. None of these things suggest good looking, but he *is* and very much in his own way. His features aren't African, except for the oiled black of his perfect skin. His features remind one of black Egyptians or perhaps Ethiopians. There is something Paleolithic about him, perhaps Basque-like, a bright primitive sense. He thinks of himself as Aboriginal, as in of a first people, as in harkening back. To look hard at him, over a camp fire light, once you got past worrying about your own safety, you might expect from him a language of clicks and pops. That is until he spoke, for then the sounds of his words surprise you with comfort. They do not match his weathered working exterior and the elastic way he moves.

Benji Aboo is a polite dwarf of great learning and it permeates the sound of his speech. More than that, he is a dwarf who has vision, poise and jurisprudence. He presents himself through his speech as smart yet humble, someone who is aware you are not his equal. He wears that burden carefully. Though he is capable of reticent shadow talk, that which holds him back in threatening situations just outside of annoyance, he is more purely himself when he speaks plainly and with graceful assurance. Hear him and you know he will treat you fairly. He is a diminutive man of enormous stature and presence. To put him in some sort of narrative context, he is not a born leader, he is a *self-made* contradiction and mystery. Benji Aboo, in another time would have been a wizard's counsel, he would have been a shoe-repairman who wrote poetry, he would have been a steam-fitter who played clarinet to himself, he would have been the roundup cook who gently trains broncs after dark when all the tough guys have fallen asleep. He would have been the night janitor on the hospital's graveyard shift who sits bedside with terminally ill children and old people, telling them in whispered singsong fashion of lives available to them forever.

He is one who has mastered the tricky business of leaving before others become dependent upon him.

And now, with this young man of such critical promise, life-tested and seasoned, Benji found himself allowing his math to slide towards fragile suspense accounts.

Benji had slowly digested what Enno had told him. He would work to be in the presence of people important to the young man's story, people like Sloop, Jackson, Victoria, Jimmy, Riven and others. He didn't need to meet them, just be in their presence when they are going about normal business. He would then be able to read their individual math. Understand how they 'fit' in the equation. Whether they fit in at all.

He already knew and loved Shirley. She was his bird of light. When she had asked him to protect Enno, he understood that it was a foundation task. And, now, he owned the responsibility and understood everything he had ever done before. Most particularly those things that tore at him.

"We are all of us the slaves of external circumstance: even at a table in some backstreet cafe, a sunny day can open up before us visions of wide fields; a shadow over the countryside can cause us to shrink inside ourselves, seeking uneasy shelter in the doorless house that is our self; and, even in the midst of daytime things, the arrival of darkness can open out, like a slowly spreading fan, a deep awareness of our need for rest."   - Fernando Pessoa

"In 1958, at the age of twelve, I spent 50 cents, a fortune, and took my eleven year old brother with me to see Elvis Presley in the new movie King Creole. Presley drove his WATCH-OUT train from the future right through the neighborhoods of maternal boredom. The black and white scenes in the New Orleans bar, with arabesque swirls to the music, those images and coming-air stay with me still. My brother never got over it and became a street corner,

manic bible thumper, fanning himself with a fist full of pamphlets and a flopping tattered little old testament. Our parents were convinced that neither of us would survive and that juvenile delinquency was an incurable disease. Ah, if only that had been true, maybe we would have grown up to be like the Everly Brothers on rum. Instead we split and took two different roads to cultural insolvency." - author.

Everybody must pervade
> *bringing songs inside the house*
Nobody wants the paint
> *to crack til earned*
There's cows and dogs
> *to think about*
all down along the curves
> *and stones boiling through the burn*
repeat yourself, repeat yourself to find
> *a way on through*
the raven's hop will
> *thank us ahead of harmony*
Everybody must pervade
> *under the cover of song.*

Chapter Nine

sour dock

restless, anxious, insatiable vanity, fed by
the gossamer fickle of fashion,
makes of today's depressed society
a cannibal's buffet.

Brahms Clarinet Quintet in B minor, Opus 115.

Trying to honor Sloop's concerns, Enno would not go into Mascara. But this trip, more than a hundred miles over the Cascade mountains for supplies, seemed fine. Duden, Aboo and Resumé were on a trek, north of Eugene to Harrisburg, to get some seed and composted chicken-manure fertilizer. When they stopped in Springfield for gas, the young attendant, asleep to the world, perked up when he saw Enno; perked up and then stopped in his tracks and just stared. Duden avoided his eyes and passed him two twenties for gas, he then went round the building to use the men's room.

Paying no attention to Benji, four foot tall black Benji, the attendant asked over his shoulder, "Who is that?"

Benji said nothing. The boy gas-jockey went into the store/office and spoke to the girl behind the cash-register, beckoning her to come out. She was annoyed but came. They stood by the gas pump talking and then Enno came round the building. Both of them went silent, both of them stared.

"Who are you?" insisted the girl. Enno frowned, then smiled softly, shook his head, looked down and moved to get into his pickup truck. He rolled up his window, stroked his dog's head and ignored the kids. Benji, got out of the truck, finished with the gas pumping and put the lid on the tank. Then he got back inside and said, "We better go now." Enno started the engine. The girl now had both hands flat on the driver's side glass staring hard at Enno. He smiled at her, rolled down the window a little and said, "We need to be going, please step back, I don't want you to get hurt."

She wheeled around and gasped. "Did you hear that?' she said.

"Yes. I told you."

Enno couldn't make out the rest of the conversation. He drove off.

The entranced attendant said, "I'm going to follow them. Shut the station. Call the others."

Shirley had completed her mission, had found her answers, and fallen to depression from the sluffing cliff of stupidities. She was the daughter of a crazy old dead hippy woman and a drunken Alaskan native. Hard to find the basis she was looking for in that. Hers was a lofty itch, she wanted some sense of where it came from. What she had discovered of her parentage felt, at best, peculiar. No loft.

She had two handfuls of mail which had been forwarded to the Dismal Nitch, OR, post office. Only one piece drew her in. It was a post card from Victoria and Sig in Guatemala. They had gotten married and setup a household in the Central American rainforest and wanted Shirley and Enno to know they would always be welcome. There was a set of cryptic instructions coded as only Sig could do. She smiled and decided right then and there to take some of the last of her D. B. Cooper money and travel to visit them. She was not in the right mind to go to Enno, not yet.

The ghost of Titus Ibid was beginning to understand in that way that ghosts digest transparent circumstance. His focus would not waver from the Duden boy. When all was well, Titus was comfortable. When something threatened the man-boy, Titus was alert in that way that tightened ghosts to excess. There was no revelation or epiphany or aha moment, it just happened that Ibid knew he was assigned, long term, to this place of guardianship. Enno Duden had to be protected, and Titus was one of those who had been given the job. The difference in his case was that he was a transparent cloud-sitter of past and future consequence. Riding, invisible, in the back of the boy's '52 five-window Chevrolet pickup, and witnessing the two crazed seekers at the gas station, Titus understood he needed to make sure those two had no success following. That he might do, but he had no way to control the news spreading.

The red-enameled Vermont Castings wood heater has that way of turning dark, burnt sienna-ed, purple-lake as the dry juniper firewood reaches full temperature and the warning label melts away. Giving off heat, warmth, warning and solace.

As I write these words, struggling to knit them together, not for you - hapless reader - but for the task of revelation, I am regularly taken back before to thoughts of stealing rides when young. I wish I had known back then...what? Wasted thoughts.

These the 'free-shipping' times, of packaged threats as opiate; they demand of any transportive storytelling effort that the platform and delivery simplify, complicate, intensify, soften, use light oils, and build around life layers of lacquered armadillo-shell casing to protect all of us from Artificial Intelligence.

< mr. author, before you get too damned serious, narrator here; I notice you've got this thing about names. Over the top I might add. Why fight it, go with the flow is what my mother always said, so I've come up with my own. Afterall, the narrator can have a name, right? So from now on I want to be known as Dude Ironymous Binge. Don't you love it? I realize that with our rules of the game this may be the only place it appears, so I wanted to make it a name with legs, one which might at any moment get up and walk off on its own. Good luck holding that back. >

Humor does not lead, nor is it a destination.
Humor is how one may receive, nose plugged.

Dr. Fil had spent his adulthood in scientific inquiry while engaged in hand to hand combat with academia and industry. He had learned in hard ways how it is that the developed world is made up of organized crime structures, of syndicates and cartels. The power companies are of a massive, loosely affiliated cartel. The big organized religions are syndicates. The auto industry is a massive cartel. The health care industries are syndicates tied to the insurance industry and serving at the teat of the pharmaceutical devils which in turn secretly buck-up the Russian and Hispanic drug cartels. Federal governments are crime syndicates tied to one another loosely as a set of massive cartels. The cyber world in all of its manifestations is an amalgam of crime fiefdoms. In every single case all of these sections of the fruit we know as human society have found themselves all-in on the same synthetic, deadly, poker round. Individuals be damned.

It is, for those at the top, all about controlling the 'covet' of gratification. Every syndicate, every cartel, only survived if it controlled its share of human covet. It may not be the end times for biological life but it certainly smells like the end times for so-called advanced human societies. The single greatest threat to all of this morass would be any semblance of a 'second coming.' The mobsters of humanity understood and feared implicitly the power of one man's example to shape all men. They would agree on the necessity to destroy any such 'example' at the earliest possible time. Knowing this Dr. Fil's insides went ice cold with complete understanding of the brown dwarf's revelation. The coincidences were startling. The young man owned the land which held the keyway to the future. And the land held the young man who was the key to any future.

Jon Drummer is the drummer for the jam band *Thorn*, and, when the corporate 'handlers' were ditchable, on occasion he would do percussion for the '*Budapesht* Philharmonia', and Sesame Street. *Thorn* has a cult follow-ing and draws enormous crowds wherever it plays. Drummer makes a lot of money with *Thorn*, doing boomshuckaboom work that is well beneath his understanding and translations of heart beat. His interests lay with Classi-cal music. The musk and timber of Ella Fitzgerald and the thick emotional thievery of the essays of Clarice Lispector.

On the rock stage, fans take extreme measures to try to get 'at' Jon Drummer, to 'get him,' to '*proj* his vibe,' to 'plow his folds,' to 'verb his rolls,' to 'cloister his noister.' But Jon keeps separate, keeps his eyes closed, keeps his brain equal to his mind, keeps his wrists reading his fingers, keeps his mumbles in mambo time. Jon Drummer keeps to himself.

Once, when under the influence of a spoiled Ice Wine buzz, he let himself go, I mean really far. He went out on stage at Red Rock, Colo-rado, all seven-foot-one of him, stark and absolutely butt-shiny naked, except for a pair of 3M™ PELTOR™ SoundTrap™ Tactical 6-S Headset MT15H67FB-01 ear muffs that had been custom encased in iridescent blown glass outers which reflected the idiotic fog-shrouded pulsing stage lights in weird and numbadelic ways. He could not hear a blessed thing, but he could feel the rhythm. So he closed his eyes and did his drumming set as had been rehearsed with the band a thousand times.

When he came close to finishing the set he opened his eyes and saw that the band was standing idle on stage watching him, and every one of the twenty thousand fans including the governor of Colorado and his children's nanny were standing with their mouths open. He dropped his drum sticks and peeled off his ear muffs. Three and a half seconds passed before the crowd went nuts, screaming, whistling, yelling, applauding, and throw-ing beverage cups. He looked all around, and to his left stood a stage hand with a long robe draped over her arm. She opened it and approached naked sweaty Jon Drummer as if he had just won the heavyweight crown.

After that experience, some called it a melt-down, Drummer went into hiding; left the band, the Philharmonia, Sesame Street, his family and drove west until he found a quiet piece of the eastern slope of the Cascade mountains where he setup a tent on the tail end of the Metolius River and fished with a stick and a worm. He left behind all that was wrong with the world. For drummer Jon the woods offered the suggestion that melding with a stream, a doing, would grant re-entry. One pleasant day he decided to take a long hike, after ten hours of walking he stumbled onto a scene with a young blonde man and an old four foot short black man building a shed in the stoney pines. A funny looking small dog barked once. He ignored the dog and walked close to the men offering, "May I help?"

> *Time was*
> *we drilled down*
> *today we*
> *dial down*
> *escape we must*
> *from the us,*
> *the numbing us*
> *of cause,*
> *squeezed from*
> *the phlegm*
> *of myriad*
> *convenience,*
> *escape to ignition,*
> *disappear*
> *separate*
> *elongate*
> *pat down*
> *the patterning*
> *dust,*
> *and park*
> *out of sight*
> *of the passing*
> *confluence.*

Beyond ahead, to those moments supposed to be explained by these.

His had been a grand and nearly perfect life of good health and cheer. Surrounded by family and friends until they were all gone. His body began to knot up and thicken until all simple things became impossible so he grew thin. When he could no longer scratch where it itched, and that moving itch was everywhere, he shut down and died. Just before he died he looked around and said to himself, 'I've seen all of this.'

This long story requires now that characterizations expand while overlapping, that the plans be somewhat exposed, as least as far as to an evidentiary path.

We would insert here these things, perhaps. If we do not keep all of the promises, it will rest on the tones of it to bury the judge's terrible impatience.

Fit in:
a. young, blustery, careless, bust-it-up energy.
b. same type but in mid-life and requiring liquor.
c. 'big jo' energy, righteous.
d. L. J. energy, slow, deliberate, and conclusive.
e. a mother's 'hold-back' til vengeance releases...
f. a dog's conflicted energy; whether to vulture or protect.
g. a child's vulnerable energy, sole purpose discovery.
h. the idiot mob's energy which only destruction may abide.

i. a scorned mother's energy.

j. the thin, brittle energy of the narcissist.

k. the professorial energy in cowardice and self-protection.

l. the gamer's walleted energy always separated by screens and clicks.

m. the editor's sorely conflicted energies.

n. the tied and bundled energies of the priest, pre- and post-ordination.

o. the super-charged, spastic energies of the manic in terrible release.

p. the horror-bound energies of the deranged struggling to give their infected spirit the release of bloody birth.

q. and the 'accepting' peace of the messenger of beauty's solace.

The steamy, verdant, mountain range of Sierra Santa Cruz is the last unprotected rainforest in Caribbean Guatemala; a place where the concept of lost is forgiven. It is of the eco-region of the Peten-Veracruz Moist Forest and is home to a secretive, over-full, capacious, verdant, inspissated, diverse, and fat trove of species, from big cats and vultures to endemic frogs and scarab beetles.

Here Sig and Victoria carefully and respectfully share their tiny forest homestead with Ocelots, Chinamacoch Stream Frogs, Golden-cheeked Warblers, King Vultures, Craugastor trachydermii, Baird's Tapirs, and Black Howler Monkeys. The plant life is staggeringly beautiful and strange to the civilized eye. Resident life is protected by its cross-haired frailty forcing a tightrope walk. Walk wrong and die.

Mr. and Mrs. Maltesta's corner is a piece of the global regions of tropical and subtropical moist broadleaf forests, thick in their five-tiered overlaps, as different from Enno's high desert juniper and pine forest as dry sand is from wet wormy leaf mold. Overfull and, because of it, threatened by man with mass habitat and species extinction.

The temperatures in this tropical forest region vary only slightly and the rainfall is gargantuan. The trees are both evergreen and semi-evergreen and their varieties number in the thousands shielding and cradling, with their explosive growth and its resultant canopy, the highest levels of species diversity on the planet. Animal life in the wildest corners of these forests is loud, beautiful, various and largely uncharted. Sounds, smells, textures, and the races of growth and decomposition make of this a constant stew of life.

Sig and Victoria Maltesta came to Sierra Santa Cruz with a mission to launch a guerilla campaign against logging and the corporate rape of this primal and essential land. With Sig's secretive ways they had some important success. With Victoria's natural ability to organize the natives, new strengths and possibilities had been found. The two of them came to love the place, the Indigenous people and each other. The place suited them also as a perfect place to hide.

It is here that their visiting friend, Shirley, would find, with the aid of Sig's manner of meditation, her own brand of hope and balance. It is here she would find, with Victoria and the natives, that giving in to faith was to turn one's self inside-out, a way to accept and to interact which granted demand as a form of forgiveness. She discovered that allowing herself to be needed and to need others gave growth to each individual. She was invited to ingest useful ritual, and in this way to make matter, and be of matters, and to matter.

"Clever ain't enough, there's got to be an idea'r in there someplace."
 - Walter Brennan as Amos McCoy

'Bout now is when I, as the author, need to hold on, because my belief in this writing - smart-assed, florid, wordy and weird as it is - is the only thing keeping the process going. A reader, after the fact - after the book is complete and away - may find that the threads of plot are enough to carry one over the swamps of introspection and stupid comedy. But those good people who have read to this point, patiently waiting for the next chapter to roll out, they must be fatigued. They must be heading for the exits. So from somewhere about here I need to go it alone, without oxygen or a net, Dizziness the new normal, every moment tomorrow, every sentence written on bad credit.

Any critic worth a damn knows he needs to get on board as an after-thought, jump on the carousel as it spins, fall backwards on to the boat as it leaves the dock, opening eyes without assignment, for the moment of opined judgement is junk and does not go well with the genuine.

<We whisper, this is not your mother's writer, this is not your brother's friend. Here you have an author whose pen is out of ink yet he flamboy-antly threatens to write more blasphemous grammar-syphilitic sentences in the obvious wish that the learned will rise up and throw him out of the bookstore.>

Henry James, "The Middle Years," 1893 *"We work in the dark — we do what we can — we give what we have. Our doubt is our passion, and our passion is our task. The rest is the madness of art."*

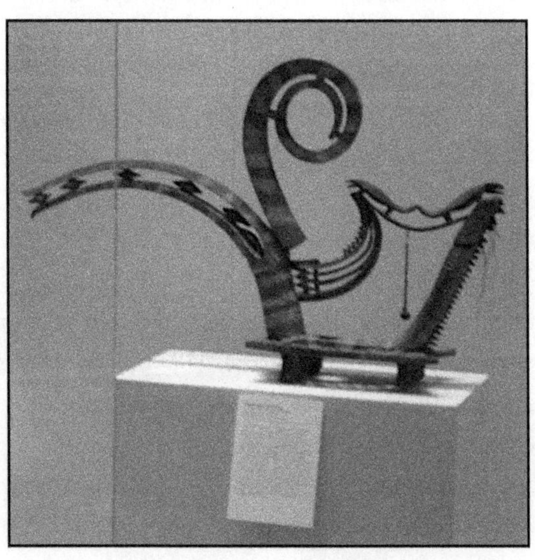

Chapter Ten

cheatgrass

"Man tries to make for himself in the fashion that suits him best a simplified and intelligible picture of the world; he then tries to some extent to substitute the cosmos of his for the world experience, and thus to overcome it. This is what the painter, the poet, the speculative philosopher, and the natural scientist do."
> *- Albert Einstein.*

"I am almosting it," Stephen Dedalus - from one of the many versions of Ulysses - perhaps by James Joyce

Humph. The lasting curative power of journey never would belong to the visionary, always rather to the small dogs, eyes close to the ground, determined to get home.

Duden, Drummer and Benji molded into a steady, joined force for construction. Enno and Aboo learned a little of Jon Drummer's strength and character as they shared the stacking of stone, mixing of mortar, and plumb checking, all without much conversation. What was to be a short foundation wall grew by the inertia of enthusiastic energy to a stub wall and then continued on to waist high; that's when they stopped to decide if it was time to set anchor bolts for a framing plate. Surprised to find so many stones of perfect shape for stacking, every rock solid in its seat especially with the mortar, the work of each next step came natural.

Benji observed, "We need the windows, or to know exactly what size they might be."

Jon said, "Thirty-eight inch holes were a safe bet for those two doors, but windows, that's something else again."

Enno stood and smiled with his eyes at the building taking shape. He said nothing. He was not crowded.

When they broke for lunch it was like he was all alone, Duden's thoughts went full force to a memory of working with Sloop as they got a field ready to seed. John B. 'Sloop' Ogdensburg, once his boss and friend, had been explaining to him how so much of the reward of farming was in witness. How he loved seeing the field tilled and curried in wait for seed. And knowing this was a temporary, transitory perfection because, with luck the seed would germinate and that same plot would turn growing colors. Then over two handfuls of weeks how it was that farmers are treated with the magnificent sight of the burgeoning crop. Each transitory stage, each changing to the next, all of this perfect sacred trust. There was the surface of it and there was the deep and valued history of it.

Watching the boy in reverie through sideways glances, Benji spoke to Drummer, "Have you done much construction? You're good with your hands."

"Nope, chasing beats is about all I've ever done." He made a motion like he was rattling a drum roll on a snare drum. Benji smiled and nodded and, to Drummer's pleasant surprise, asked nothing further. Yessiree, Drummer was gonna enjoy hanging here for a while.

After witnessing horrors between Guatemalan government forces, American mercenaries, and rainforest natives - feeling helpless and exhausted trying to make sense of the mind-numbing, senseless, rampant death and disfigurement folded in to a swirl of swamp mysticism and caucasian stupidity (whole 'nuther book in that) - Shirley kissed the Maltestas goodbye, reconnected with her meditative self and headed north to find Benji and her Enno.

She may have been in a better, if goofy, head space but, alas, the de-

tached daze made her a mighty easy target for professionals in search of
Duden - she didn't care about things close at hand, like people tailing her,
mentally she was already in the Oregon pine desert, she was already pos-
sessed of forward thoughts, those flailing, limp, sore, smooth and deciduous
hours in Enno's arms.

So professionals like Leo Vitch, real name Leonid Runaroundavitch - a
dangerous man on loan from the Russian secret service to help the (Off)
White House keep America freshly fascist, focused his eagle eye on the
young woman's route. This potatoey guy was a treble agent. Besides the executive branch of the formerly great U. S. of A., Leo was being paid by the
Myrmecophilus Group, a consortium of organized religions and high tech
companies concerned about emerging symptoms of individualism threatening the doctrines of fevered shopping carts and tithe-funded ceremonial
rituals.

And Leo Vitch was also a fixer for the GeeWhiz corporation.

But the most important professional operative in the hunt, The Monk,
recently retired and married, was a whole 'nuther piece of the puzzle. He
remained, even out of practise, generally considered as the premiere hit-man
cum-terrorist on the planet, certainly the most feared. Sig kept pace with
the underground information network. It helped him and his wife to stay
beneath any radar. But when they received word of a highly lucrative death
warrant out on an Oregon man named Enno Duden he was drawn in. The
Monk, aka Sig Maltesta, told his wife Vicky she would need to hold down
the farm. Sig needed to get Vicky's maybe-brother, Enno, out of harm's
way.

"If you hurry you can catch Shirley. If she isn't helping you, she will
hinder. And honey, see if you can bring him here for awhile? Why not both
of them?"

Don't know how he did it but our ghost Titus Ibid managed to sit in on
Drummer's thoughts, Shirley's musings and Sig's concerns all at the same
time. He ran through a number of moves on his mental chess board and

threw out all the consequences but two. These were outcomes he needed to
help arrange.

*"Presumption is a deal breaker. I won't have it in my house of business.
There may be no such thing as certainty I demand, but one way to guarantee
that is to make critical decisions based on casual presumptions. Presumptions set
the table for paranoia, erratic behavior, even ego-mania and you see where that
has taken our governance. And before any gawd-damned one of you nods your
head yes, remember, I expect steely-eyed stoicism at these meetings. You were
not selected for this board to be sheep, you were selected to be cynics and worse.
Profits must follow, yes, but way back there where we need binoculars to count
them. Our vision comes first. Today, if we aren't at the forefront of efforts to
save the planet this corporation will shrivel and die. Traditional business models
are for losers without talent. See the future, good or bad, and believe in it.
NO government on Earth can begin to compete with or control the far-sighted
corporation unafraid of well-dressed maximum effect, maximum market share,
and maximum share value. The only thing we cannot compete with is new-
found faith and hopefulness for the common people. Yes, you heard me right.*

*"'We cannot compete with hopefulness. And, in today's reality, we are at
risk IF some messianic force of purest form raises its ugly head. All bets are
off if another Christ, Che, Buddha, or Mohammed gets captured in someone's
phone video and goes viral. All bets are off if that happens. And he or she will
not come out as a center-stage activist or political entity, nope. Instead, our
researchers have run the algorithms and what we will likely be faced with is
yet another barefoot innocent mingling in sparse crowds or small groups, a long
way from power centers and cities. We've got early 'heat' readings on a couple of
possible "saviors" out there and professionals have been contracted to neutralize
the threat.*

*"Now, on a related subject, Jason Weened has a report on Gloryholes, our
initial work to combine sports arenas with churches and pharmaceutical distribution centers."*

Not a single member of that board of directors flinched, they all trained their tight, mean, unseeing eyes on the CEO and hoped their bladders would hold till lunch time. Another form of 'yessir.'

Dr. Fil thought about Dr. Goodenough's glass battery and felt kinship. Significant others were working on better ways forward, just as they have for thousands of years. Which brought him round to thinking about the little man, Benji Aboo.

Dr. Fil's intelligence quotient was off the charts because his thinking was off the charts. As a youngster of seven and eight, his dad had dressed him in girl's clothes and pushed him in the direction of fashion design. Some parents will go overboard grooming a young child for a certain sport or musical instrument, Dr. Fil's dad wanted his son to go in the direction he had denied himself.

When little Fil was ten he would have to sneak off, to closets or bathrooms, for excitement and fulfillment. There he would read chemical equations, and complex theorems on collapsing econometrics. The much encouraged silly intrigues of early onset fashion curiosity had, even while he fought it all the way, given the boy nearly limitless horizons of frilled thought; how the body form and the garment form could talk with one another and come to compromises which were absolute enhancement. He saw ideas fully clothed yet transparent. Now, as a successful if renegade scientist, a career his father saw as a patented failure, he could appreciate what the little man had shared with him. And Benji, sensing a mind that could handle it, had given the good scientist quite a boatload of information.

Benji's tale:

"And, throughout time, including today's, the self-possessed, the designers, all of them, went in search of leaders. It had come to this, that no rules or formulas or type were called upon. To be among the designers was to arrive at instinct's command. The leaders, those naturally so, would stand out, away, and immune until their time. They would learn to slip in and out of superficial guise, slip around and through crowds, avoid connection - avoid friending - avoid absolution - avoid questions - avoid buffets - scrums, locker rooms, rehearsals, testimony, contracts, celebrations, housewarmings, fish frys, arena crowds, attorneys and accountants. And to do all of this without being branded as curious. For a natural leader to stand apart, that is a challenge.

"I offer an example: here was a young Swazi chieftain, a boy, really only barely a man. He had been chosen by the designers of his people. The stories came to him of the World War II Nazi invasion of North Africa. Dead of night, he quietly left his village and walked hundreds of miles to volunteer to help rid his continent of that hideous scourge. Later, after years of war, he would sneak back to his village - older and permanently altered - to be designated once again as it's divine leader - now, with obvious, though mysterious, new strength - now, with the nursing visage, the holding touch, the calloused calm. Years later, when dots were connected and outside folks realized this old chieftain had been instrumental in saving thousands of lives during WWII, all at great risk to his person, he was decorated for his bravery; almost an insult after a long life of selfless service to the deep needs of good humanity.

"It all started with an unresolved young man surprised to be selected as chieftain and never certain why, not yet self-possessed. Applauded, decorated, anointed too soon, before the proof."

Dr. Fil adored his collection of manual pencil sharpeners, those little ones you twisted with two fingers. He used a different one each day. Carried them in the coin pocket of his Carhartts. He bought cases of graphite pencils, ordinary and exotic, and kept them in a supply closet which doubled as a large fire-proof safe. He also stored in there his scientific papers. Plus he had hand corn shellers, special application pocket knives, hand-crank grinders and milling devices. He did this for the art of it, he did this for the joy. It flew in the face of the stuffed-shirt cloistering of scientific rigor and posture. He did not do this as a regressive survival tactic but to be ready on the outside chance that mankind might live on and thrive and need alternatives to electronic convenience. It was his sideline.

His hyper-intelligent abuse of electronics is legendary in certain circles. He holds the domain of convenience-tricked electronica, that human endeavor/dependency, as equal to the arrogance of genetic engineering, the deadly devilry of chemical farming, the usurious graft of payday lending, the prolapsed intestine that is political engineering and those dirty diapers that are the nature of most churchifying.

But in Benji's tale he saw antidote and positive skeletal structures. Dr. Fil was so taken by Benji's last caution, told slow and with eyes closed, that he wrote it down to read over and over again.

"I surprise myself forward for the long night eats as the longer day drags - pushes for the trials for beauty extending beyond life to next life, and the ridden living of life.

"Today with man's law spinning to abandon, anything that may be seen as an effort, posture, or outsider's decoration may be actionable. It is not a question of whether you will be found guilty of nothing in particular but rather will you be one of the very few that never is.

"Nature's laws as opposed to the laws of men: no fool is above man's law, most fools are in perfect harmony with nature's laws. The clever enfranchised man who lives in spite of or in constant battle with the laws of nature is mostly above man's law, lonely perhaps. For him nature is the terrible equalizer. But these clever are horribly trapped. Their lives turn inward towards decay. Oh, but if they could only be foolish. The only answer is to save yourself, get out of every political and religious argument so that you may be heard at the trial for nature. Be a witness not a jury member. And dress carefully for in many countries of the world capital punishment is banned for all crimes except impertinence.

"You see Doctor, genius may no longer be an acceptable excuse for misbehavior. The law has lost its languor."

Many people slip in and out of the 'zone'. Duden is always in the zone. He hasn't a clue what it would be like to be out of the zone. When he takes a shot, chooses an action, it goes in, which is not to say he is casual about it. To the contrary, because he knows it will go in, he takes extra precaution to make sure it is the shot, the action, he wants to take. Very few times has he made the mistake of allowing emotion to rob him of his careful decision making and some of those times have resulted in great harm. It is why he seldom speaks without slow rumination, which frequently results in the choice to say nothing.

The fatigue of depression stole him. Sloop was limp with it. Helga came to the same conclusion but hers was reactive. She could not abide by her husband's manic phases, not any longer. His depressions saddened her to compound low gear.

In his brain he crawled toward the one thing which might save him, again. Beauty, the sort which requires immersion. Farming and ranching had taught him this as had his friend Shoulders, by example. Sit with it until it owns you and it will cut the grease of depression. Helga did not know this about Sloop. Married enough years that they grew their own interiors without sharing, without witness, without transfer.

At the core Sloop believed in himself so he freely borrowed of himself, certain the debt would be forgiven, yet fearing that the end of life audit would turn up a failure.

Still and all, he had his posture which sent undeniable messages.

How beautiful might anything be which comes from the hand of man or woman? And, why must someone decide ahead of merchandizing? And, who may claim the authority to judge after?

Beauty was purpose
  and the other way round
The outside wants we
  explain horrors and loss
the inside must have
  purpose of beauty
I've tried so hard
  to fit in
and it is killing me.
  Man-kind is
outside of itself
  always every time
The one with pain
  is inside
The pain is unnecessary.
  No humanity.
The blood is done.

Circus acts: where the frightened fighter enters the crowd-engulfed ring to *demon*-strate his ability to neutralize an adversary only to leave exhausted and empty. The knife thrower's lament. The premise behind the painfully exquisite French film "The Girl on the Bridge." (The knife thrower saves a lovely young lady from suicide only to convince her that serving as his target for a carnival act would convert the waste of suicide to exciting purpose. Of course, love must enter and twitch the knife thrower's wrist.)

Well spoken anarchists wearing designer socks the colors of old bandages.

In his curative dreams Uncle Nimbo runs on the Decentonian Party ticket for County Designate, duties of which are knife sharpener, match-maker, mollifier, and apologist. His campaign slogan is 'let us laugh and cry until lots of good is done'. He always wakes to the reality of his large fat form and slobbering visage to laugh off standing for public service. After breakfast he breaks into a sweat and, with time, comes out of 'it' fine.

A mob formed that decided to take justice in its own hands only to find that just over the hill another larger mob had done the same thing only to wrest justice from the first mob as a third mob organized by a parent company came to swallow the second mob while a fourth mob guided by a crazed attorney general, guided in turn by a crazed president, guided also in turn by polls read by a really big company and interpreted into fourth grade me-me English which was spoon fed through television broadcast to the oval office, which by the way has become so ovular it threatens to become lima-bean-shaped.

Jeb, the hit man, came out of the hardware store to find the dusty Volks-wagen Passat he had rented with a note under the wiper, flapping in the stiff wind. It read, "Time for you to leave." Oh that made him mad. So he decided he would leave, but first he would finish the job.

Titus made a note in the way ghosts make notes.

Orson Welles is said to have favored the Negroni cocktail remarking that the bitters were excellent for the liver while the gin was generally bad, all in all a good balance. The decidedly American rationale of 'offset.'

Spoilage in humans, manifest in that rank, absolute, bitter and destructive entitlement posture, that of the dead-hearted self-possessed, has nearly always resulted in shellac-brittle fake idols - only slightly elevated on their own dung and surrounded by a shadow moat of a grease slick. The meek know there aren't any clams or edible roots 'neath that slick, only the shattered glass of dribble crystal.

'You want me? Then come and get me!' she screams, but no one comes - not any more. The souls of the spoiled are encased in mirror.

Ah, but the souls of the suffering masses are quite another matter. They aren't encased in anything. Each a lonely soft pearl made of broken gratitudes. In a long, lovely chain they hover over the collected pains and sorrows as if some designate rope of rhythm fog 'round the long neck of life. The Hungarian viola player can be felt, as much as heard crying, as he searches Lewis' percussions for *A Milanese Story,* to soundtrack that beautiful rope of gelatin pearls as it waves in a curvilinear dance to broken-hearted Macedonian dreams.

Christ's announcement, long misinterpreted, did not, as the translators insist, declare that these meek shall inherit that earth. He proclaimed that THIS Earth's only chance for survival rests in the heart of the meek for they are the embodiment of the careful passenger poise and a laborer's love. What he did not say is what he assumed would be understood: the meek are waiting to be shown what to do. They are waiting for the one person who does not want to lead, who does not want to control, who does not want to be followed. They are waiting for the one new virgin soul, that person with no previous incarnations, the one person who is immune to society, the one person who God has chosen to leave to his own devices.

> *The gods offer no rewards for intellect. There was never*
> *one yet that showed any interest in it...*
> *- Mark Twain's Notebook*

Of the forms old as song

Chapter Eleven

nightshade

*"When you break the great laws, you do not get liberty,
you do not even get anarchy. You get the small laws."
- G. K. Chesterton, 1905*

*"If you plan on breaking a big law,
best you have its replacement well in hand."
- Titus Ibid*

He splashed olive oil across the large, older, cast iron skillet and placed in it one whole chopped yellow onion and seven chopped garlic cloves. Once hot he laid in the three large, one and a half inch thick, bone-in, pork chops with salt, pepper, white wine, French mustard, pomegranate-infused dried cranberries, one small sliced apple, a touch of Balsamic vinegar, six splashes of Worcestershire sauce, one splash of Soy sauce and a long thin line of squiggling Blackstrap molasses mimicking conga-lines of ants drunk from a spill of Kahlua. A wide cover was placed over the pan with room to breath and the chops were flipped every ten minutes. It all cooked this way for 40 minutes. The results were exquisite and will be remembered for the remainder of his days, much like that cherry-glazed duck they enjoyed at the discovered Vermont back-woods inn thirty years ago.

Then his brain went limp.

How the old brain slips so easily to the moron's rest area, waiting to tell itself urgencies have passed. Middle of a working day the aging man wants only to have the project move along with no requirement for decisiveness. It is early in the morning that urgent thought accesses him, propels him, worries him forward. It is late of a restful evening when his manner and aptitudes might accept he do surgery. The big chunk remainder of the day he is built to rest by the side of an untraveled road and hum apologies to nature, his and hers.

"Pepper me you warm man, now!"

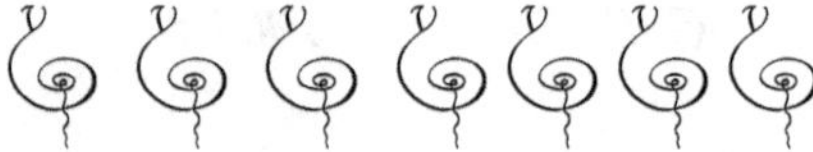

The first time it happened they were, all of them, sitting scattered about on rocks and logs in a horseshoe bend, leaning forward gazing at the building they had worked on for so long. There was Enno of course, and Benji, Drummer, Uncle Nimbo with his three big dogs, Jimmy Three Trees, Dr. Filipiano, Resumé, and, way up in the tree, Titus the ghost with his cougar friend. That's six flesh and blood men, one ghost of a man, four dogs and a tired old mountain lion - the only female. This was the group who first came to understand Enno, to love him and to fear his purpose.

They had worked a long, hard day and shared a meal. Now they were sharing a quiet moment in the dusk until Jimmy started in...

"...but that's okay, it feels like rolling, and even up a hill, because because because this box house of sticks was was was wasn't here before and now it's it's it's it's here, and don't we love it because it's a place to back into until he says we have to to to to go home, but this is home and I don't want

want want want to go and feel bad I felt good to to to today, feeling good pushing sticky cement, feel good making this fa fa fa farm for Enno."

"Yes, my shaking, stuttering friend." responded Dr. Fil, "it does feel good to work hard for no other reason than to lend a hand and make new friends."

"Doing good with no thought of being paid is one of those things people do not want to admit they value, messes with the system. The 'Man' don't like it," offered Drummer. "I played drums once for a high school rock band at their Prom. The regular drummer got the measles and they had no options. I got a call from this famous writer I know whose nephew was desperate to find a drummer. Kurt called me and asked what I would charge to sit in for the Prom. I told him our band's corporate management figured that my time, including their cut, was fifty-thou a night. No way to afford that. But more I thought about it more I wanted to help but, you see, I couldn't let on who I was cuz our corporate handlers and bosses would go ape-shit worrying about losing control and percentages and I'd have to pay their share out of my pocket and then fines on top of that. Big mess. But I did it - snuck in and did it, and for free. It was way fun. Though, even with my disguise, eventually the kids figured it out. When you're a famous stick man over seven foot tall, hard to escape those phone cameras. Luckily I wore that wig and mask. Damn that was fun."

Uncle Nimbo said, as respectfully as a recluse might, that he didn't understand people paying a week's wages to hear loud music. And then to pretend they were drunk or high on something and jump around, wobbling their heads side to side, and moaning between their screams.

And Drummer said, "You got that right, man. And you say you haven't been to one? Even back when you were young? The Stones, The Kinks, The Finks, Mott The Hoople? None of those great bands from a hundred years ago?"

And Dr. Fil interjected, "Hold on son, the seventies and eighties *aren't* a hundred years ago."

"No offense. I know it isn't that far back but it seems like it."

And Jimmy said, "I like drums."

And Benji offered, "Some cultures use drums and drumming to communicate across long distances."

And Jimmy said, "I like ice cream too. No- no- no- nobody far away gonna take my my my my ice cream."

No way to trace back to where the understanding began, but it was clear that within this group of new friends no one pried into the personal business of the others. Drummer had found himself volunteering information about his other life. And Nimbo wanted to talk about his dogs, and Dr. Fil was always keen to explain his take on theoretical physics.

They all talked back and forth and over each other, except for Enno who listened, gazed at his cabin taking shape and slowly nodded. Benji said a few things and watched Enno. Of course Titus, as a ghost, said nothing.

There came a lull in the conversation then a quiet spell.

Enno spoke. His voice was calm, certain, and soothing. But a surprise nonetheless.

"I want to do good. That's my purpose, and sometimes it confuses me. Still I want to do good. I decided for me that means farming. I want to be a good farmer and do good farming. I must do it with my own two hands."

It was the most any of them had heard Enno say unprompted. And the clarity and simplicity of it, along with the solemn tone, held them quiet for the rest of the evening. They were in the presence of something less and they wanted more of it.

Heads down in prayer posture, they forbid time to pass until time pulled at them to break up.

Uncle Nimbo and his giant dogs, Mikey, de Gaulle and Sarkozi, went to their trailer and whittled sticks thinking on what they felt.

Drummer went to his tent thinking on what he felt, knowing he needed to stay on longer.

Resumé and Benji went, with Enno, to Riven's cabin and thought about what Enno had said.

Jimmy and Dr. Fil went back to the motel rooms and thought about what they felt. Jimmy felt a warm wet like he had peed the bed. Dr. Fil was excited because, though not completely understanding it, he knew the pieces were coming together with new definitions of wholistic.

Titus laid down beside the she-cougar, smiled and slept the transparent sleep of ghosts on task.

Shirley, a thousand miles away, thought on how she felt.

Vicky, further yet, tried not to worry about Sig loose, angry and determined. Then hugged how she felt.

Houston and Bessie lay shuddering in each other's arms thinking of how they felt.

Cash Apercu counted his money again and knew how he felt.

Lloyd burnished the forged chunk of iron he intended to incorporate into his sculpture of the 'purposeful man', and let his brain waft.

Vic Nib kept sticking pins in his thigh hoping just maybe he would feel something deeper than this silly pain.

The much maligned idiot president of the U.S. snuck out through the tunnel to meet up with the old fortune teller/retired porn star on the fourth floor of the old post office building.

Sloop and Helga sat around the little kitchen table saying nothing to each other and having a deep and complete conversation nonetheless.

And Enno Albert Duden thought about how he might make connected enclosures around the perimeter of his garden for a small flock of chickens. And how that allowed they have different grazing each day and helped to keep down the bugs. And whether or not he should plan on having a Cao de Gado dog like Uncle Nimbos' to sleep in the stone-lined enclosure with his sheep band. And how he wanted to talk to Nimbo about learning how to shear sheep.

As Sloop struggled to protect himself from the deadening depression, to stay engaged and positive, his body began to reject itself with rashes, flaming itches, pistules, new hard crusted warts, muscle failures, stutters, fogged eyes, ringing ears, and wandering pains. It was at these times his reason left him and the resigned sloth of self destruction set itself to new ritual. It is the body in conversation with depression. Once we gather ourselves together, gather good thoughts, we realize there can be no good conversation between body and depression. It must cease absolutely and finally.

"The mind is not, I know, a highway but a temple, and its doors should not be carelessly left open."
                    - Margaret Fuller 1844

Benji's thoughts: There is gold in the doing. Our heroes grow smaller as we grow older. They shrink before our eyes, not that they diminish but rather that they get shorter with age and the blinding harms that come of *our* petty wisdoms. They get so small near the end that, if we are lucky, they disappear behind our measurements of their accomplishments. So we begin anew, apologizing for them by 'discovery'. The grand works they did; if we are further fortunate we will see the short sightedness in that and realize the perpetual energy of nature is the gold. It is not just what they did but the nature of what they did, the nature of their doing which sustains.

Continuity is a fragile keep. Left to it, it does not shove the push, heave the lift, insert the planting, dose the watering, peel the flaking. Do not excuse the death, witness without the walk-away. Closure is not work, it is cowardice. Work is the keeping.

Unconsciously he had long ago perfected the furtive touch, perfected the timing to fit in, as concentration and preamble occupied the target, as comfort no matter how transitory had snuck in only then, looking away. Would a soft palm or curled finger barely grace the target arm or shoulder followed by the tucked apologetic head? At once worrisome and endearing. The eyed sense would guide.

If you cannot see that which is eating you, cannot feel yourself being eaten, are you any less eaten?

> *Admit your fears*
> *watch shadows clump and sluff*
> *Rushdie's arrogance*
> *Dirt's moment in the flood*
> *Stay it isn't sew*
> *We weed to find good plants*

This final mistake has been this electronic rational for ultimate disregard.

robes of beauty
bode the duties
trees the pattern
says the queen
if he had two
he'd leave me alone

Chapter Twelve

lupine

The world is too much with us, late and soon,
Getting and spending, we lay waste our powers
Little we see in nature that is ours,
We have given our hearts away, a sordid boon!
- William Wordsworth

Would panic be but that layer of essential which never was?

This writing jumps around. It just does. Life jumps around. It just does. Are you all in with the reading? Are you all in with the living?

To have an agent, someone who aggressively represents, sells and places your work, this is an insidious goal for purposes will cross. But such an agency might be accomplished by the talent itself, by the individual. Delacroix, Degas, Melville, Monet, Twain, Matisse and Picasso did such things. Certainly also true within many other disciplines. To do it for oneself is to eliminate most cross purposes but it does require we wear the coat of third person which embodies a certain meanness of spirit. It requires a possessive objectivity.

Dr. Filipiano understood this implicitly. His purposes tucked deep in the purse of his psyche, he translated his agency for the time being into an advisory position behind Benji and the invisible others. The good old Dr.'s job of the moment was to protect Duden, whether the young man liked it or not. By doing so Dr. Fil would be protecting the discovery and source of Brown Dwarf particulates and perhaps, no make that definitely, doing humanity a service.

Realizations keep insisting themselves on Lloyd. The ways in which the failings of commerce insists itself on us all. How repairmen continue to disappear without replacement. How schools no longer 'muster' and how art is less artful with every breath conjoined. How the insidiousness of credit has each climbing upon his or her own ashes. How the very irrelevance of irrelevance is the blubber of blubbering, the fat of fatty, the stupid of stupidity, the false insouciance of indifference. Peace of mind versus peace of minds?

In the sunshine, repairing fence, he listened on his truck radio to exquisite choral music sung by the choir of Trinity College at Cambridge. Spell binding. Then he went back to the house with notes and attempted to find the same music on the computer. Did so - but didn't, because the quality of the sound was hideous - and it made him realize and remember. Time was, as far back as the hifi period of the 60's in San Francisco, that the quality of sound systems lifted everyone. It wasn't just the question of needles in grooves versus tapes and digital - it was speaker systems and wires. The experience back then lifted everyone. And now? When had he bought into the change? When he left at the landfill that which worked sublimely only to be replaced by the convenient lap top computer which never ever had a skin to shed or a tear that bled, which ran either on a batteryless battery or braidless cord? With his fountain pens he needed occasionally to refill them. With his 'sound system' he needed to flip the record. With his old wood cook range he needed to keep the fire going and check the oven temperature. He made himself a promise right then. He would find an old wired sound system then set it up in his studio and enjoy it and the return it represented. Along with Stacy Kent singing French love songs, Bill Evans teasing Miles Davis to improper counterpoint, Glen Gould squeezing Bach to sublimation, Astor Piazzola's bandoneon stroking the backside of a million female knees and elbows, he would find a vinyl recording of the choir at Trinity College and protect them all as purest gold.

He thinks: everyone should be left to their own 'devices', in his case, cattle, horses, old trucks and tractors, old sound systems, collections of bolts and iron scraps, and small tight and tightening communities. He nodded to himself agreeing that the world needed lift - shared lift - otherwise the buzzards of commerce would win.

What part to play in the shared lift?

He left his house and went to the studio. There, from the cupboard, he took out a can of Johnson's paste wax. On the pedestal stood his newest sculpture, *'The Purposeful Man.'* He rubbed the wax into the dry, thirsty surface of the forged iron. Later he would polish with a car-wax buffer and finally he would abrade portions of the surface with dirty sand. He had already built a shipping container for the piece. He put each sculpture in its own container and stacked these outside the studio. He wasn't done with this one just yet, but he was done being immersed in it. He wanted it to go away, out of view, for a good little while. He wanted, some confused day in the future, to deliberately happen upon it and see it anew. See if it worked. See if it drew him in and asked him to stay.

Day after day of Italian skies and cooling heat had their effect on Sloop and Helga. Depressions gone, together they worked on ranch doings, she with her garden and chickens, he with his Percheron horses and Shorthorn cattle. And naturally, they found themselves bumping into each other, asking for the odd helping hand to lift or review or share a moment. At one point, lost in a comforting pause as he gazed across the fields, she came to him and hugged his arm like a child, kissing his exposed elbow. They both smiled without looking at each other.

When we match love, work and sustenance to perpetuity we have a kitchening of life.

Dr. Fil had explained to Enno Duden, without too much scientific detail, that deep beneath Snake Flats he had discovered a massive cavern fed by a confluence of lava tubes. He theorized that the atmospheric and barometric conditions were perfect down there to perpetually incubate microscopic particulate matter in tiny reverberate explosions.

"When you are down there and sit quietly on its edge, you begin to see that it is a universe of its own. Fluorescent lint, that's the best way I can describe it. Some slightly bigger pieces pulsing like fireflies, slightly smaller ones in various colors spinning around the fireflies. And thousands of these firefly systems. It's like looking in on a miniature universe with solar systems floating in a small outer space. I tried once to walk into it and I couldn't. I lost my bearings and started to float. And my head felt like it would explode.

"I took down some jars with me and managed to capture spinning fuzzy particles on the outside edge of the system. Sometimes, these fuzzy particles would bust apart in tiny explosions resulting in a ball of millions of even smaller pieces spinning into itself. I have been studying these particles in my lab and learned that they may be a profound new source of energy and anti-energy. I have registered the name Filenium to identify them as a new primary element, but they are actually an energetic system rather than an element.

"Enno, under your land I believe is an unlimited and sustainable source of new energy which could be 'farmed' and used to heat, light, and power this planet forever. That entire cavern is a Brown Dwarf, a kind of prolapsed starship as it were. This could mean, Enno, that petroleum products and coal would no longer need to be used. That global warming might be reversed. This is so critically powerful a discovery for humanity and the planet that it is essential that it not be monetized. Imagine what oil and coal companies would do to try to destroy this and prevent it from ever being released? In the short run, the only thing which will protect this discovery from the destructive force of commerce is to keep its source from being found out. The long run is trickier but I have some ideas."

Enno had listened quietly while he watched Benji's face. Enno was naturally drawn to discussions of discovery, of understanding, of the mathematical fit of this to that. All the while Fil had spoken, Benji Aboo had sat silently and with his eyes closed. Enno had sat up when Fil spoke of farming Filenium. That flame for farming. And Benji had sighed when he heard mention of the Brown Dwarf. Dr. Filipiano's ultimate pause had caused Enno to turn his gaze and Benji to open his eyes. The only one to overtly acknowledge what the good doctor had revealed was the ghost Titus who nodded in that annoying way ghosts have. They, after all, can be such know-it-alls.

Later, Jimmy had first given Dr. Fil the idea when, in one of his stuttering soliloquies he had spurted 'when turtles fly.' Dr. Fil made inquiries and found a suitable Chinese manufacturer who could make heavy, four inch wide, pot-metal, toy turtles with bobbing heads and tiny wings. The bottoms of the four feet were coated in velcro. These little turtles were hollow. He ordered a dozen samples shipped to him in Oregon so he could fill them with Filenium. The result, after some fine tuning, was a toy turtle that hovered while its head bobbed and its little wings fluttered. With a leash and velcro-suitable landing pad, voila, *Spurtles the Flying Turtle*. He took the necessary steps to conceal the true nature of the Filenium by injecting it into Bounce lint sheets. Fil had something to introduce to toy manufacturers for possible licensing.

The idea was to keep the hover source an absolute secret and hold on to the manufacturing. He would sell the limited rights to the big toy company, Wowza, and in that way have two things: capital to proceed with Filenium research and development, plus an exciting way to introduce one aspect of the element to the general public. It escaped him that he was breaking his own rule. He was monetizing Filenium. But that would be later.

Duden wasn't really in on the deal just yet. He wouldn't accept the plan until after he was told by two people about the boiling river of Peru.

"Now is a good time to talk to you about the water. I know you've noticed that recently, I say just after the earthquake, you started to see rivulets of fresh water on your land where none had been before. And Benji told me you found their source, or at least where its coming out of the ground? In my explorations down in that cavern under your land, it is obvious that there have been big changes in the geological structures. Deep down I came upon a pool of water that was rising, albeit very slowly. Leaning against the wall of the cavern I noticed it was warm to the touch and I heard something. Putting my ear to that stone surface I could clearly hear water rushing. Easy to guess it is warm water. Running water, pressure and heat - an educated guess would say that more changes are coming and soon."

Both Enno and Benji were glued to Fil's words. Benji interrupted,

"I am reminded of the *Boiling River* of Peru. A shaman friend recently took me there to witness what is to his people a most holy place. The earth forces hot water from its bowels, up and into the passing river, bringing a four mile stretch of that wide water to a cooking temperature of 200 degrees."

"Yes," said Fil, "a similar thing seems to be starting here but without a passing river to intercept. Enno has found cold water, I found hot perhaps too hot, but they may mix to form a brand new natural hot spring of perfect temperature."

"And water to irrigate plants and pastures with?" asked Enno.

"Yes, my boy, oh yes. I think its time to find the lawyer, Ashwan Clevalure, and get a patent filed on your soon-to-be river. Be thinking of a name."

Arm in arm, he resolute in his good looks and fashion certainty, she rail straight, completely unforgiving and cold in her leadership - only as pretty as nature allowed snuck through. She held his arm to grant stinging costly chance. Everyone watched as they crossed the club floor. Houston Tuttle and Bessie Quince were relieved and exhausted as they sat in the corner booth of the undeniable LA night club. They immediately whispered interruptions back and forth while they tickled each others finger tips. In the background a laughing Jeff Goldblum and the Mildred Snitzer Orchestra successfully perfumed old standards with naughty asides which succeeded in levitating Goldblum, that rail-thin, movie-star-cum-jazz-pianist-cum-reluctant-philosopher, away from the mirrored worships of stardom. He was reclined in full view on a lithe, carbonated, tight-circle carpet ride, as though entertaining, by winks and small undulating laughs, his bevy of children as they endured a deadly storm.

A long ways away, in a sparsely populated edge of deniable Central Oregon, Enno left Resumé with Benji and drove his truck to pick up some tools. It was 90 degrees with a bright, chalky yellow air. The unusually thin dust whisked aside quickly and there were patches of visibility. He came around the corner of the gravel road, at the bottom of the incline, to see a figure ahead of him on a bicycle. It was an older man, tight and elastic with cycling muscle, wearing a stretchy bright-colored juggler's pant suit. The bike looked to be expensive and new.

This cyclist was struggling, raising up to push down with each side's pedal stroke, and letting - no forcing - the bike to wander from side to side, middle of the road. Then he rose up one last time, hesitated and fell over sideways. Enno stopped his truck and got out.

The man was dead. And he had a broad, inappropriate smile on his face. A few minutes later, as Enno stood over the body, a pickup stopped and the female passenger hollered out. "What's going on? Has there been an accident?"

Enno answered, "No, the man just died."

She rolled up her window and the truck drove off quickly.

Another vehicle showed up. "What's going on?" the young man asked.

"Man died."

"I got a cell phone, I'll call 911 and get the police here."

"No emergency. He's dead. He died happy." Enno added.

The young man rolled up his window and locked his door.

Enno sat in his truck and waited.

How Titus had reached Benji and Dr. Filipiano we aren't to know, but he did. And they drove up in the Dr.'s old van. Dr. Fil quickly assessed the scene as Enno explained to Aboo, "He was riding up the hill and it was difficult. Then he fell off the bicycle and died. He died happy."

Benji looked at Enno with his piercing brow-hooded eyes.

"He earned his passage," offered Duden.

That was all the confirmation that Benji needed. The knot was tied in that instance.

"You must go now. Take your dog with you. Leave us to handle this," Benji insisted.

As Enno drove off with Resumé, a police car sped up, lights flashing.

Dr. Fil was waiting, mentally prepared for the minor shell game. Benji went to sit in the van and wait out the clean-ups. His life's work had come to clearest definition.

This messiah was without message, he was the spiritual cotter key which held the mechanism together. His gift wasn't teaching, it was example, it was the trail his purpose would leave from each answered moment to the next. His gift was the pattern his actions showed to every unhappy slave to tragedy and boredom. His gift was nucleic and atomic and musical. Their job wasn't to seek him, wasn't to follow him, it was to absorb him. Enno was to be absorbed.

So they lined up.

The first eleven were given to protect - not him - but the destructive others, protect them from themselves for their assignments to destroy him were in actuality their own death sentences / warrants.

The second eleven were given to translate and avoid any and all efforts to codify.

The third eleven were charged to replicate.

The fourth, fifth, sixth and seventh eleven were charged to assure that he and she become usefully invisible, wonderfully long lived, and constantly light giving.

And Titus and the dwarf were to brother it all.

Many came to insist, with little sound, that the young man was 'already' a mysterious triumphant mix of opposite ingredients, a chutney-like spiritual condiment in quiet personality form. When all he actually was - that steamed guava-like primal essence - came of the miracle of cyclone carried seed caught and squished together in an island palm frond and 'containered' for future plantation.

If there was a bright slice of hopeful light flashing repeatedly from 1770 to 1820, when science and music and politics and painting stood up for everyone and everything, and futures, hard - beautiful - merciless - sporadic - ascending - lubricating - delightful - HOPEFUL - futures, hid laughing at every corner, we our fathers let it all go to waste in our godless pursuit of dead perfections - of self-oiling profit - of perfect unkillable enemies - of wholesale misguided mercies disguised as understanding - of corporate protection and aggrandizement, of the circular high.

Fifty years in the so-called modern history of man, when glimpses came every day of how things might be and how terrible things were. To have both views handed to us with God whispering "isn't it obvious what is to be chosen?" And we went giggling with the whores of commerce down through darkest alleys to where our blood was sucked from us as we signed contracts granting our souls be fuel for the monstrous corporate hounds.

These be the life ending nightmares of Abe Lincoln and every other leader who ever gasped at the ledge of humanity's compromise.

The fight with inevitability is lost unless we are willing to see in ourselves the stuff of the obvious villains of the day. The smallest, quietest, lovely heroes in our midst have always been the light forward and for forward. Their self-evidence so obvious in invisibility, we felt compelled to apologize for the awkwardness they brought to the altars of commerce.

Those heroes in our midst understand that to poison the future of any life form or biological spectrum is evil to the nth degree. It is a form of biological genocide. By comparison murder is far the lighter crime, heinous as it be. Yet aren't we all complicit as our government, in our name, proceeds with 'naturecide' every day and all for commerce?

"All that the senses can but imperfectly comprehend, all that is most awful in such romantic scenes of nature, may become a source of enjoyment to man by opening a wide field to the creative powers of his imagination. Impressions change with the varying movements of the mind and we are led by a happy illusion to believe that we receive from the external world that with which we have ourselves invested it."
  - Alexander von Humboldt Cosmos 1845

Give me the splendid silent sun with all his beams full-dazzling,
Give me juicy autumnal fruit ripe and red from the orchard,
Give me a field where the unmow'd grass grows,
Give me an arbor, give me the trellis'd grape,
Give me the fresh corn and wheat, give me serene-moving animals reaching content...

 - Walt Whitman Parade Music 1865

Tygh's lunch menu:
Smoked goose breast,
King Bolete mushroon
Smoked Asia pork roast marinated in Hoisin
  and Sriracha with rice wine vinegar.
Enchilda. Mole over smoked goat leg stuffed
  with garlic and serrano peppers. Marinated overnight
  in cilantro pesto.
Oregon razor clams in green sauce - an enchilada
  with corn, brown rice, peas with smoked broth.

Sparkle Donkey Tequila
Whistle Pig Rye
Aylesbury Duck Vodka
Piehole Whiskey
Lubricating the reception of
a soaring Jacob Banks
and Charles Bradley's primal scream

"*A browndwarfish quote maybe. Seems almost an orderly syllogistic form, this whole thing of being able to think in front of a tiger. Versus behind a tiger's back, or aboard a tiger? Starting with the point of balance, as if with a sword or spear.*"
 - *Paul Hunter*

"*If you know the point of balance, you can settle the details. If you can settle the details, you can stop running around. Your mind will become calm. If your mind becomes calm, you can think in front of a tiger. If you can think in front of a tiger, you will surely succeed.*" – *Mencius*

Part Three

Concha

Chapter Thirteen

thistle

Who has not found the heaven below
Will fail of it above
God's residence is next to mine
His furniture is love.

- Emily Dickinson 1860

Enno woke up with a start. Across the inside of his brain was the command, "get ready for the water." He knew what it meant but hadn't a clue how the thought came to him. He also knew in that instance that the coming waters would be called *the Sourse*. And he asked himself "how do I know this word?" and he felt the answer this time without rejection. He knew that it came before, some place some time before, and that it belonged to him and that he would know more, some place some time he would know more.

*mid-14c., "support, base," from Old French sourse
"a rising, beginning, fountainhead of a river or stream"*

And then, fitfully, he went back to sleep. What woke him next, and completely, was the sound of his own voice saying,

"I don't see any useful answers in what you want to do. None that rise above. Nature always offers answers that don't require questions."

Titus weeped.

Young Enno understood. If you are to survive nature, you must first have a lifetime of living within nature, listening, smelling, touching, marveling, cursing, praising, even exalting nature. From this will come an acuity which may, at that fateful moment, allow you to hear the whispered "duck."

Old Titus felt himself in the presence of... and then he felt himself flashing in and out of being. Was he done? Had he finished his assignment? Before he greyed out he felt himself argue that this was not gardening.

Aren't these those things, those things mothers know?

And where have the women gone? Where are the women in this story? Why do we ignore them, fear them, discount them?

Vicky in the rain forest

Shirley enroute

Bessie Quince

Penny?

Helga?

Nettie

Olivia

Ashwan Clevalure's lost love

Time to introduce a bad woman. She comes to the story as an insert.

Nowhere is the name, Ulrika, found in the chart of characters, on the lips of the regular characters, as a warm wind in any of their histories, as siren, as vixen, nor welder nor graveyard shift assistant manager. She is the nasty top of the food chain. And she is without emotion unless you consider insatiable hunger to be an emotion. For her, logic is a staircase long ago worn to submission. Ulrika never had an obvious source of income, never needed it. Whatever she wanted she took, but only if it weren't offered to her first. Ulrika Freya Orly is a ruler of wolves and as such first one to partake. She is incapable of true cruelty or actual kindness, she is matter of fact. It would seem she must be Scandinavian but no one knows, not even her, for she has no memory or recorded history before her young adulthood. Her persona is so electric and changeable that no one who knows her can describe her. Her hair color changes from blue black to chestnut to champagne. He eyes are colorless, which is to say they reflect the color before them and so are any color. Her skin is molten and honey-toned. Her neck is so long that with certain clothing it appears her head hovers over an accompanying lanky body. Her lips may be perfect or crippled in their twist, snarling or sweetly enticing. Her fingers are popsicle sticks, her elbows are gator rough, her knees are soft as warm butter, and she rides behind a skinny beast of a man on an old, ridiculous, small 1968 250 cc Benelli Cafe Racer. She and her man lead a motorcycle wolf pack of tattooed and disillusioned dentists, accountants and lawyers that call themselves The Heaven Reapers. Most of them wear wigs. Her man's name is Tronk and Ulrika has no appetite for Tronk with or without his wig. Rather, she has an appetite for enabling him. She captures or entices women for Tronk and the rest of the Reapers. When the 'gangs-hers', as they call themselves, are done and tired, she hypnotizes the women and sets them loose at laundromats. Then she tells the men individually, when they wake, that the women left because they were disgusted with Tronk and his Reapers; that the boys had done a poor job of their work. He cries, they cry and she feels as fulfilled as she can without emotion. She is that motherless, childless, needy mother we all fear, the one who haunts our sleep. She is the queen of wean, she is gas-fired dart, she is like long-overcooked spinach rolled in boiled zucchini and sprinkled with tabasco.

Tronk and Ulrika are crossing Idaho headed for Oregon on their way to their favorite hunting ground, Seattle, bath tub to the rich and insouciant. Ulrika likes Seattle because most of the women don't require hypnosis. Oddly, they avoid Portland because Tronk is mortally afraid of Voodoo Donuts. When Tronk is particularly troublesome Ulrika threatens him with an old, soiled, Voodoo Donut bag. Make no mistake about this, Ulrika and Tronk are not hip, they are the poison mankind has long deserved. If the vernacular is the result of congestions of small brained people struggling to identify through tortured, stilted, familiar language, imagine that the Heaven Reapers eat the vernacular.

Dr. Fil sat across from Clevalure the attorney, a short, wide man looking all the world like a Manhattan store manager of double-breasted Baltic origins. Yet, urban as he appeared, he smelled of forest, wet rock and thick, slow-simmered sauces. He spoke in low, measured, back seat tones, eyes without blink, while slowly rolling his yellow pencil between his fingers.

"Such a claim must do more than simply state the law of nature while adding the words 'apply it.' Generally speaking a new natural phenomenon is not patentable, and that includes any shielded argument of location specifics as pertains to deeded land. There are curious exceptions at play as regard claims of copyright for naturally occurring items if it can be shown as critical parentage to new forms. The precedents are still in question and have applied almost exclusively to genetic engineering. That would certainly not apply to a new stream or river. I believe what you have here is a simple case of water rights, unless I am missing something else?

"But please allow me to change the focus of this discussion for a minute. I was lead to believe that you are speaking on behalf of a young man of limited resources and one who might be taken advantage of by larger forces. Is that accurate?"

"Yes, Mr. Duden," answered Dr. Fil, "is quite young and has nothing except his mortgaged interest in this land. To speed things up, let me guess your next question. What interest do I have, and how might I benefit from helping him to secure his claim in the new waters? I hope to secure from the boy a share of the mineral rights of his property. If that land were to be subject to any uncertainty, stemming from the perceived value of the new, hot, artesian spring and the river it is creating..."

"Did you say hot spring?"

"Yes..."

"That would quadruple any value of that water for both government and private sector. There are steps that could be taken to obviate any effort to claim this land by eminent domain, and perhaps even to protect it against some commercial piracy. I will need to do some research but action should be taken as quickly as possible. May I reach this young man directly, hopefully this afternoon? Have you any issue with that? I want it understood that I will be representing him, not you."

We are the cowardly aggressors, we must try everything, no matter the expense to others, but the bargain is that once we are done trying everything we must shut up about it, otherwise our last act will be suicide, not humiliation. And not because of our crimes and debauchery but because we are incapable of facing our emotional insolvency. Because we do not want to see and hear how empty we are.

On the radio the commentator spoke of an effort to redefine homelessness to include folks who were renting and unable to buy a home. Something about expanding definitions of hardship to include inconvenience and entitlement thwarted.

It took Houston Tuttle less than a week to shake off the death of his most recent lady love, Bessie Quince. For him her death had snapped a trance. Bessie had been struck by a driverless taxi in New York City. But before that she had succeeded, in short order, to instill in Houston the idea that his charisma had power. Bizarre, with her gone he immediately and naturally thought he wanted to use that power for good - though he was equally certain she had other ideas. He was ready for a change. Houston Tuttle had recently been asked, as a joke, to consider running, as boy candy, on the Decentonian party ticket for governor of the city state of Chicago. Instead he leapfrogged to put his name forward, two years in advance, as candidate for president.

He had discovered boutique vodka and lavender bath salts, the results of which compelled him to make of his constant travel and public speaking a campaign for celebration, pleasantness, and ruthless agnosticism. He did not want to be president, he wanted the campaigning. He wanted a stated purpose for his public presentations. He wanted the soap box because he had a mission. Simply put, he wanted every poor, sad, depressed and crippled person to find happiness. His campaign slogan was "a smile in every heart, a planet on the mend." He invited street musicians to write songs for his campaign and he took two or three different ones with him to each rally or stop.

The rag-tag red-headed beggar, with French horn - fist in the bell, played a soaring rendition of *Carnivale* while a short round baritone bum sang the new words...

We are the hope of the world,
    though we fill these holes with despair
        every smile reaches up
            every child reaches up
                clueless and wanting to share.

He is the hope of the world
    Houston Tuttle knows how to care
        we will follow him
            as he follows us
                clueless and wanting to share

He's our country's best hope, he's our planet's best hope,
     he's our family's best hope
          for repair.

The crowds at the rallies took to singing along and swaying with the
music. The rallies also included circus acts on the stage wings as Tuttle
spoke, and women of color reciting poetry and reading ethnic food recipes.
It was all a show, an ever-changing and mostly low-key show, with flourish-
es and somber moments. Early on, when serious types would insist that this
all wasn't useful and that Tuttle needed to talk policy, he would tell short
parables that deflected and endeared. And when weirdos and extremists
attempted to interrupt, the crowd swallowed them and spit them out the
back. Soon it was all about being at the rally and enjoying Houston Tuttle,
up there, entertaining, reassuring, leading. It no longer needed saying, they
just wanted Tuttle to tell them it would be ok, and he did so immediately
and often. He showed them.

At the beginning and end of each rally the crowd, in unison, would raise
a hand and together say "I will." From both sides it was mesmerizing and
dangerous. But Tuttle understood and deflected that threat of devolution
in a repeated comic act of hot potato juggling. He'd holler out "remember
otters come from otter space" and then he would wander out and into the
midst of the crowd turning, with sincere naiveté, to see who might climb
to the stage next. Nobody did, nobody could imagine going on stage after
Tuttle. The crowd hugged him, but at a safe distance of 37 inches.

Later and only slightly jaded, Houston Tuttle soon found himself calling for the formation of a shadow executive cabinet made up of musicians,
magicians, jugglers and poets. In the beginning no one seriously thought
about voting for him. They gave him small contributions, lots of them. So
he kept doing it until all of a sudden the crowds grew to tens of thousands
and the major media took notice. Lots and lots of money and such noise,
and fervor. With time the Democratic and Republican candidates for president withdrew. No one wanted to challenge Houston Tuttle (except, under

cover of the slime cloud known as corporations).

Smelling perfumed blood the journalists went after him, more so because he ignored them completely. Instead of answering their demanding questions, he gave his very short, non-story to genuine writers, thinkers, poets, and community organizers. So journalists came up with what they thought was an "aha, wake-up you dopes" moment with the oft repeated headline question "What if Tuttle?" It backfired. People loved the catch phrase and it appeared everywhere, on billboards, t-shirts, taxicabs, buses, on lawn signs. The words started to join, blend and meld. "What if Tuttle? A smile in every heart, a planet on the mend." That's when the assassination attempts began. And that sealed the deal.

Every three or four days there would be another dramatic failed attempt to kill Houston Tuttle. Having no policy points to discuss, the media turned to analyzing these ugly shootings, car accidents, bombings and what have you, speculating on who was behind it. Houston's followers understood that government and business were behind it which only further validated their candidate, so they worked through social networking and guardian-angel policing to protect Houston and take the offense to the hit-men and psychos. But more important, they organized into pods of hyper-intellects with a mission to infiltrate government and business decision making and turn the fear of Tuttle to a *We are Tuttle*.

In time Houston Tuttle would go on to win the presidential election. And most everybody was disappointed. They didn't want the campaign to end. Least of all Houston. So he appointed two women and a man to act as a presidential *witenagemot,* a pre-parliment if you will. And Tuttle, copying a successful tactic of a former idiot president, continued to stage rallies.

Filson Wool is a finish carpenter and premature geezer. He was sitting in his pickup truck reading something into a cell phone.

"Men are nuts and women are fruit. With women you've got ripe ones and you've got hungry ones, you need to see and know the difference. Ripe can sometimes be hungry, not often. But hungry is seldom ripe. And hungry is often destructive. It's man's great handicap that he cannot see the difference. Men are never ripe, not like young women. They can be young, succulent and attractive but that does not make them ripe, as in fruit. They don't fairly demand to be picked and eaten, as do young women. But men are almost always hungry, and as such destructive." He looked up and out at nothing and he continued. "Of course there are the deviant pseudo genders to throw a wrench into it all. But it doesn't have to be hard to understand them. The lesbian is basically a hand grenade disguised as a man. And the gay is basically a cream-filled pastry disguised as a woman. Now, I am not one of those who see this as a new phenomenon, no siree. Been around since time immemorial. Difference is that in the beginning they were considered fair game, then we morphed into a time when you had to get a license to hunt them, and now they are protected, not as an endangered specie, 'cause lord knows there are more of them than us, no now they are protected because the lawyers are losing sleep trying to decide if their own gender is morphing. Now I know there will be some buttwipes who will find my remarks hurtful and perhaps even incendiary. May I strongly suggest that those folks need to find gainful employment. For, if truth be told, mankind has always been an invasive species which is long past threatening the entire planet. As such, humanity in all its variant nastiness is, or should be, fair game for those of us who insist on thumping the toad's brother, so to speak."

He sat and nodded his head into the cellphone. He stopped nodding. Then, as if for us, he provided one side of the conversation.

"I don't see what's so funny."

silence

"It's because the truth isn't supposed to be funny."

silence

"She left ME, the witch. But I still love her, can't help it."

silence

"No, I'm not using it right now. It's behind the old fridge in the shop. Pull it out from the left side. And close the damn door when you leave!" He held the cell phone away from his head and poked at it with one finger until he was satisfied he'd turned it off, remembering all the while how it was that you used to just hang up your phone.

He went back to his old aluminum drift boat and the two open tubes of J-B Weld, glad that it was only a 22 caliber hole.

Filson Wool is not important to this story, so you don't need to remember him. He's like the equivalent of stinking human mulch, shading the understories, conserving the emotional moistures, providing hiding places for slugs - worms - snakes - and empty beer cans while rotting a little himself, rotting right back into the granular, essential nastiness of humanity. The only salvation for Filson - 'earlyrot' - Wool was for the right woman to accidentally force herself upon his sorry composting carcass in an obvious act of self-lubricating abandon. That woman comes up a little later. Her name is Olivia Perchance.

Oblivious to his pursuit, always one short step ahead of the Monk, Shirley crossed the de-militarized zone that separates the U.S. from Mexico and found her pickup truck and trailer right where she left it. Awakened uneasily by the stark metallic reality of the Marshall law flavor at U.S. borderlands, Shirley came out of her daze and finally sensed that something else, in her own radius, was wrong. Out of the corner of each eye she was sure she saw two men, a hundred feet apart, intent on her movements. Then, to the right she made out the shadowy form of her friend Sig, rapidly closing in on one of her tails. Taking this as her cue she turned and walked directly toward the other man, the one on her left. He fidgeted and turned, quickly disappearing through a maze of dumpsters.

When she reached Sig, he was standing over a prone figure and going through a wallet.

"Hey. He'll be okay and he'll know what hit him. Back of his hand is my calling card. This one's a fed, means things are getting very thick"

Shirley bent over and looked at the cooling red wax blob with the 'M' shaped trident mark pressed into it.

"Was that wise? Thought you wanted to disappear?"

"That can wait, there's a contract out on your boyfriend. We need to extract him right away and it will take both of us."

They rode together in her truck and Shirley told the Monk about Benji Aboo and Uncle Nimbo. Monk told her about the contract.

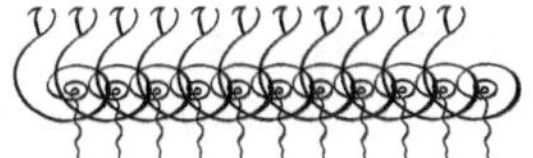

The wreck of the Byzantine ship, the Gaspaccio, was discovered in deep, dark waters of the Black Sea and included a wrought iron codex to an ancient astronomical divination quadrant or computer, what moderns call the *Antikythera mechanism**. It also contained a pile of barnacle-encrusted Budlight cans and a pristine bag of styrofoam cups with Bob's Big Boy Burger symbols stamped in highway paint on their sides. The styrofoam cups were absolutely pristine, and certainly reusable, even though they had probably been down there, a mile deep, for 25 actual years, time travel notwithstanding.

All those centuries ago, in 324 AD, it had been ordered by the Byzantine emperor, Constantine I, that the Gaspaccio, a many-oared barque, test the new Aemos theory of astronomical guidance and computation-aided navigation with a stolen mechanical astro-plotter. They were to chase a pesky ghost ship (the 21st century rental yacht, Giselle) and capture the bikini clad sirens seen basking on that modern ship's deck. Constantine I wanted to personally inspect the clothing articles for absorbency and light refraction.

> *The antikythera mechanism is thought to be one of the most complicated antiques in existence. At the beginning of the 20th century, divers off the island of Antikythera came across this clocklike mechanism, which is thought to be at least 2,000 years old, in the wreckage of a cargo ship. The device was very thin and made of bronze. It was mounted in a wooden frame and had more than 2,000 characters inscribed all over it. Though nearly 95 percent of these have been deciphered by experts, there has not been a publication of the full text of the inscription.*
>
> *Today it is believed that this instrument was a kind of mechanical analog computer used to calculate the movements of stars and planets in astronomy. It has been estimated that the antikythera mechanism was built around 87 B.C and was lost in 76 B.C. No one has any idea about why or how it came to be on that ill-fated cargo ship. The ship was Roman though the antikythera mechanism was developed in Greece. One theory suggests that the reason it came to be on the Roman ship could be because the instrument was among the spoils of war garnered by then Roman emperor Julius Caesar.*
>
> *X-rays of the device have indicated that there are at least 30 different gears present in it."*
> *- from somewhere deep in the internet where*
> *authors fear their works are mistaken for trash.*

Jon Drummer had met a limping old man on the Pacific Crest Trail. After visiting a while, the man declared that he had come to a realization his life was all but over. He offered Drummer three hundred dollars in cash if he would take his fishing pole and tackle box to Cultus Lake, southwards in the Cascades, fish with it, and then leave the gear high in the elbow of a certain curling Pine tree. The tree was at the far northern end of the row of cabins, near the cul de sac, and had the words "it was ever just us" carved into the barkless section of the trunk. Jon refused the money but insisted that the old guy tell him what all of it was about. So came this story:

"I know it now, didn't fully realize it when she was still alive. We had thirty years together yet it took her absence for me to understand, my wife and I were so perfect for each other because we were so combatively different. If I would declare something to be so, she would curl in thinly disguised disgust and make a valiant effort to ignore it all. When, on another occasion, she would bubble over with anger and insist that this or that was dead wrong, I would storm off in silence. She had her friends, I had mine. We seldom drove anywhere together, she'd drive her car and I would take my pickup. Knowing such things you might think we were candidate to split up. But it weren't gonna happen. We were just too much in love.

She loved to fish off the banks at high lakes. I'd go with her and maybe fish for awhile but mostly just hang out and watch. She did not like to eat fish. I did. So I'd clean them, she'd cook and I'd eat. Our own version of Jack Sprat and his wife. On our last trip together..."

At that point the old guy had a stroke. Didn't kill him but twisted him horribly and he couldn't speak. It took Drummer the better part of the day to get someone with a phone to call for an AirLife helicopter. As they were loading him on the stretcher he grabbed Jon's sleeve and pointed to the fishing tackle. Jon nodded yes.

Drummer did the deed. Never heard anything more about the old fella. Thought about Schubert's unfinished.

"In the brighter light of the seaward side of the house
she looked extremely well, her face tanned and smooth. On
her mouth she wore poppy-colored lipstick, and gold mesh
jewelry on her arm, a heavy gold chain on her neck. She had
aged a little - she must be thirty-eight or thirty-nine, was his
guess, but her dark, close-set eyes, which gave her a fluid and
merged gaze (she had a delicate, lovely nose), were clearer
than he had ever seen them. She was in the time of life when
the later action of heredity begins, the blemishes of ancestors
appear - a spot, or the deepening of wrinkles, at first increasing a woman's beauty. Death, the artist, very slow, putting
in his first touches...

Ideas that depopulate the world."

- Saul Bellow Herzog

Chapter Fourteen

locoweed

Where this time wants us to go in order we survive.

'No object can be tied down to any one sort of reality; a stone may be part of a wall, a piece of sculpture, a lethal weapon, a pebble on a beach… Everything is subject to metamorphoses.'
- George Braque 1905

Benji, Resumé and Enno arrived at Snake Flats very early that brisk fall morning, before any of the others, and entered into a thick, fluffing, warm fog. The deeper they went in the more it became apparent that this was steam. The smells were akin to a sage-scented shower stall. Benji made mental notes about the elevation of the lower edges of the fog, the bottom of the exposed root-spray bursting out from the ancient Juniper which had grown up against a leaning flat rock, a foot up into the cluster of Aspens. He was relieved to see that the cabin that had been built was above the edge.

Enno was thinking about what Sloop had told him regarding the Andean terrace farming of the Incans; how it was that a thin slip of fog rising from the water in the shallow ditches at the edges of the individual terraces kept the potato plants from freezing. The young man was wondering if this steam-caused fog could be diverted and used inside a greenhouse.

Benji took out his notebook and sketched a layout of the farmstead they had discussed, suggesting to Duden that they could use the level again as a make-do transit to shoot some grades and determine if there was a suitable way to run a diversion ditch, flat and off to the side of the artesian source.

Enno went to his truck and got his longest level and make-shift transit tripod. Then he got his wide tape measure. When they were done they had a meandering line of stakes marked to grade.

They both noticed that in a couple of days this new source of water had already begun a process of carving the land. Sandy loose top soil was being washed down new paths. Enno tried something simple. He started to stack a row of stones across the path of the new water and noticed immediately how some of the debris-rich soil slowed, filling around the rocks and forming an earthen dam which in turn spread the waters wider. Watching this Benji recognized the young man's instant translation of phenomena to action. He wondered how they might do these same things on a larger scale. That was when they heard Jimmy approaching.

On that sharp promontory of rimrock, Titus stroked the head of the old cougar and listened as it coughed. He was not where he belonged, yet so much depended on him being right there. From her desk at the library, Penny - eyes closed - could see Titus. His right half was black and white as it should be, his left though was a palette of muted colors that fluttered as though in a breeze. She knew something was going to happen, she felt several skies cracking open.

So much occurs in ridiculous and provoking overlap. While Enno was working to start a farm from scratch, Dr. Bone Stock was attending a Paiute breech birth, Houston Tuttle was running for president with scratch, Shirley and Sig were designing offense, Senator David Actual Roger was orchestrating a scratchy campaign for Oregon to secede from the union, and Lloyd Shoulders was making travel plans to go to New York City for creative combat.

Trusting his instincts, Ashwan had changed from his double-breasted three piece suit into woods-wear. Dr. Fil had offered that Jimmy Three Trees might take the attorney directly to Enno at Snake Flats. Fil stayed behind to work in his lab, so he missed the transformation. Jimmy, while with Ashwan, did not stutter. What he said still teased the laws of sequence, but there was no stutter.

"This boy is mine and he won't be for long, soon he's a man, maybe now cuz the light hides his edges. His sister went away and I didn't. Has nothing to do with water just jumping out of the ground. I'm glad he has a good dog. I have good dogs. Sometimes I have good thoughts, big ones, then later they are gone. That ever happen to you?"

Ashwan nodded solemnly as they entered the steaming fog. Jimmy hollered out for Enno and Resumé barked.

The first one to come into clear view inside the fog was Benji. Seeing each other, Ashwan and Benji both smiled, at first tentatively, then wide.

All that changed when tall Enno stepped through into Ashwan's circle of vision. A rush of adrenalin surged through the little man and he found himself lowering his head to the young man. It was Benji who helped with the moment,

"My name is Benjamin Aboo and this is Enno Duden. You have a purpose here." Not a question, a clear statement of the obvious.

Ashwan explained he was an attorney and that Dr. Fil had retained him to determine the best way to secure the new waters.

"I came here to meet you and get a fuller picture of the situation. Now I understand it is not just about the new waters, it is about everything."

Duden turned to Benji, and with that Enno walked back into the deepest fog with Resumé and Jimmy on his heels.

Clevalure and Aboo sat on stones for hours and talked easily from end to beginning.

When it came time to leave, Ashwan found himself remembering the old Wilkens boys, Jimmy and Bobo. They were batchelor brothers farming in a deep narrow valley of the coast and they had been his clients. They were gone now but they had left him as trustee of their estate which included their "Cabinet of Reading" plus a barn full of painstakingly cared for farm implements. Bobo had said, often, "when we are gone you will come upon someone and it will be most obvious to you. You will give him or her the cabinet and the implements along with these notes on how to care for the implements. Might not seem like much but it could make the world a better place."

The Wilkens boys had been quirky and more than a little mystical. Ashwan had trusted their wishes. Now, having heard from Benji of Enno's purpose, and having felt for himself Enno's presence, it was clear to the judge/attorney where the inheritance was to go.

The Wilkens boys had allowed, early on, that a church be built on a corner of their farmland. Back in the forties, it had thrived for five years then the congregation had moved on. Jimmy and Bobo were avid readers, mostly of obscure literature and farming texts. They collected books, new and old. They built tall shelving and filled the little chapel to its high ceiling with those volumes. Long ago they had heard of the Portuguese Cabinet of Reading in Brazil and took to referring to their tall library/chapel in the same way. They saw this as the treasure it was and provided a fund in trust that it be either maintained there or, if need be, moved to another suitable farm when the "heir" was found. The chapel was to come with instructions.

Ashwan had told Benji of this and they had agreed to get Duden there to see it for himself. Meanwhile Ashwan Clevalure, pursuant to Enno's wishes, had drafted a patent on the *River Sourse* and began the process of securing the waters in the public domain for perpetuity with Duden and his choice of heirs as stewards.

Titus heard all of this and was thrilled. The cougar coughed.

Nimbo and his dogs, camped as they were on the furthest corner of Duden's land, were making their rounds to check on the sheep. Mikey, DeGaule and Sarkozi were a bit distracted this morning. They couldn't get it out of their collective heads; that little dog, the one called Resumé, had actually been chewing gum and blowing bubbles. Mikey, at one point, had actually tried to push open Resumé's mouth to see inside, to sniff what was being chewed so lovingly and gently, only to be rudely surprised when a sticky bubble inflated and burst into his face. The big dog yelped, then growled and sat back. In that way that certain dogs do, Resumé smiled to the first edge of laughter. The other big dogs, DeGaulle and Sarkozi, sat and watched, heads cocked sideways, fascinated. They needed to learn this. This was something they wanted to share with the boss. It was these shadows on their brains which had distracted them just enough to miss the approach of the man with the rifle. He had gotten inside their cordon, which meant they now actually surrounded him. When DeGaulle finally got wind he let out two sharp yelps, the signal to Nimbo, followed by a sustained low growl. Immediately Mikey joined in with a midrange growl, and to complete a perfect and frightening harmony, Sarkozi added his high-pitched growl. Vic Nib froze in his tracks and slowly, methodically, lifted his rifle. The flash of Nimbo's signal pistol and the rain of red powder caused him to cover his eyes with one hand and in that instant the dogs had him.

From a distance Nimbo watched and flashed a hand motion. The dogs looked his way and saw the signal. Mikey drug the rifle over to his boss while his buddies each grabbed one of Nib's legs and drug him away, his head bouncing on the rocks, Vic soon lost consciousness. When he came around, hours later, he was laying adjacent to his pickup truck, both legs bloody. The pain was excruciating, he loved it. "Now this," he thought, "this is worthy, this is good."

Drummer was leaving the convenience store when the old biker troop pulled into the parking lot. Tronk pushed past him grinning stupidly and Ulrika slid back and forth on the motorcycle seat as though polishing it with her buttocks. It was the first time Jon Drummer felt the clear separation. He no longer wanted this world of indiscriminate appetites. This social world gone wholesale and vulgar. He wanted to be back at Snake Flats, sorting rocks and feeling the perfume of nighttime Elk farts whilst he watched the old dog chew his gum. He wanted to choose his retail and make it small. He left right away to return to Snake Flats.

Ashwan, the lawyer/priest, had pulled off the road and walked down to Squaw Creek. Sitting between the lush greens and the dry reds he bowed his head and filled himself slowly with the strength of new certainty.

Sig and Shirley stopped in Barstow, He wanted her to know about the safe house he maintained in the back of the Habitat for Humanity complex. There, in a closet full of computers he rapidly established a lacework of emails, pinging in and out of federal agencies, suggesting that hit men know they are being pursued by government operatives and are pretending not to. And that a right-wing cabal was well along in setting up an elaborate "cultural" sting against the FBI, with video feeds streaming out to Russia, China, both Koreas, Pakistan, the office of a certain failing Scottish golf course, and Kamala Harris' bath house. He was building a circular firing squad.

Sig explained to Shirley; Never make of your enemy a martyr, always work to make of him either a grateful ally or a frightened fool - preferably a fool as they will always keep you on your toes.

All the world is not a stage, most of it is made up of parking lots, waiting rooms, warehouses and crowded empty spaces any of which might be a stage but stages do not exist without the nasty and delicious presumptions of 'enactment' and 'posturings' and 'fainting' spells as announcement. Stages, in spite of Ruskin's insistence, require those hollows that separate audience from actor, they are never galleries. Serious authors might be crammed into tiny safe alley corners, miniscule galleries, oily salons, so long as they resist membership to the dead zones of commerce. How can we know who we are unless we allow that great chunks of our inner swirls remain with us? We are not what we toss out but what we retain. We are purely 'of' our content. And our content would be our contentment, finally. Integument. Gertrude Stein be damned.

We find ourselves granted
the taken for taken from
envelope edge glue dry before sticking
clothes wrong for accident
wrong for the lights
we smile lied to by strangers
we anger looked down upon

<Writers have squeaky voices and stutter a lot. Narrators have sonorous tones and are capable of wave riding the best structured sentence to even higher impact. Writers often see through their own inadequacies. Narrators never do. Writers work in their pajamas while narrators prefer Tommy Bahama. Who would you rather drink with?>

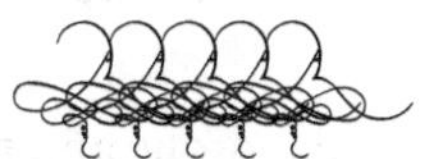

A few days later, Enno and Jackson go to an auction.

They rode in Jackson's 1980 Ford crew cab pickup pulling a gooseneck flatbed trailer at least 20 feet long. In the back of the truck there were three plastic creamery crates. One was full of ropes, one of chains, and the third of tie-down straps. There was a small clamp-on winch, two cable come-alongs, and a heavy steel tool box. Bolted to the inside sidewall of the truck box was a rusty handyman jack.

Brenda, "Yummy," Fork had packed them a cooler full of fried chicken, flatbread, a log of guava paste, a mango, apples, two Look bars and mason jars of water. Resumé rode in the back seat and rested his head on the cooler lid. Every time Jackson looked in the rear view mirror he smiled and chuckled. He wouldn't trade with Duden, he loved his two redbone hounds, but he sure liked this silly old cow dog. From the very first there was always something about him that was familiar, even comfortable. Titus had said he thought Resumé had been a failed clown in a past life. "If Buster Keaton came back as a dog it would fit."

It was a crisp morning. They pulled into the "parking"- marked pasture. The sign said "Dogs must be on leash." It was Jackson who thought of it. Came natural. He took a length of baling twine and tied it to Enno's belt loop. He made a big knot on the other end and offered it to Resumé. "You are going to have to hold on to this or they will make us lock you in the truck." Resumé took the knot in his teeth without hesitation. Both men smiled, each with their own understanding. Resumé practised dropping the knot and then grabbing it again.

At the trailer they stood in line to register for bidder's numbers. When big, black, handsome Jackson Fork stepped up with his identification the three women at the table stopped talking and nervously looked at each other. Enno, hidden from view stepped to the side of Jackson and put his hand on his friend's shoulder. Seamlessly, no words exchanged, the bidder's numbers were registered.

They could hear the auctioneer's drone and the ringmen yelping. A crowd was gathered around a flatbed wagon piled with tools, electric cords, buckets, shovels and miscellaneous items. Jackson pointed to a long roll of fencing materials and they headed that way.

Stretched out across the four acres were lines of machinery, building materials, farm equipment, vehicles and irrigation pipe. The first row they got to had items stacked on individual pallets. The first pallet had barrel pumps and pieces, the second had boxes of nuts and bolts, the third had four coils of logging choker cables.

This was Enno's first auction and he fought the thrill. So much stuff. So much useful stuff. Every few minutes Jackson would lean over and point out something, offering explanations, valuations, curiosities. For Duden this was one of the most exciting version of schooling he could imagine.

The young impatient hawk
flying low deliberate
over the pasture
Little songbirds in green grass
full attention to the heifer's graze
Magpie on the resting bull's hump
buzzards and crows languishing
mother and daughter coyotes
wide circles round the wild turkeys,
killdeer with limping flight and faux cries
drawing anything away
from camouflaged nests

To owe to the world nothing
and everything
To allow and to be allowed.
To leave your parents behind
that you may become a parent
And then to gather your parentage to you
that you may exalt them
in your and their new depth

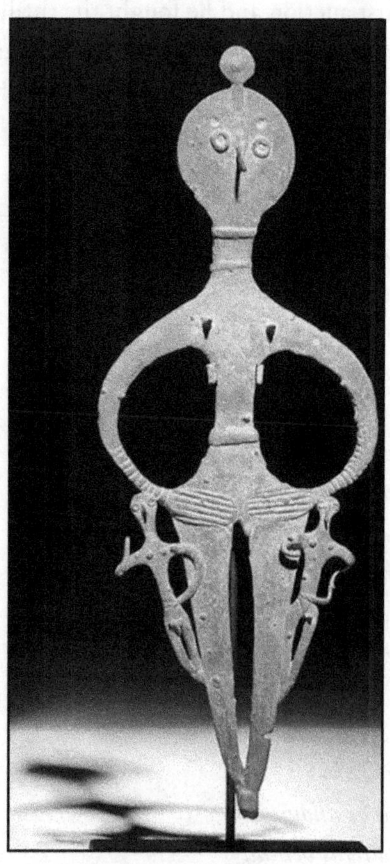

"one must love life before loving its meaning."
- Dostoyevsky

Chapter Fifteen

wet agate

"I want the following word: splendor, splendor is fruit in all its succulence, fruit without sadness. I want vast distances. My savage intuition of myself."
  - Clarice Lispector, The Stream of Life

"I have a remedy against thirst, quite contrary to that which is good against the biting of a mad dog. Keep running after a dog, and he will never bite you; drink always before the thirst, and it will never come upon you."
  - François Rabelais, Gargantua and Pantagruel, Book 1

The discoveries compounded. Dr. Fil's lab work included metered videos taken in the cavern beneath Enno's land. In this, his first chance to analyse these, he had witnessed an unexpected repeating physical process. As water droplets passed through the brown dwarf particulate atmosphere they spun in crazy yet oddly consistent patterns. He noticed that a floating pattern of Filenium would, when invaded by long thin water drips, wind that water into a funnel which increased in speed until it then wound itself into a larger secondary funnel both of which spun upside down, looking like dancing corded stalactites. And off to the side, in the video image, he noticed a stalagmite which bent and moved as though a thin tree teased by a breeze.

"Oh Geeze!" Fil's mouth and eyes, pointed upward, were the same pulsing wide red-lined circumference. His brain raced to apply a thousand theorems to what he was seeing.

The painter, Edgar Degas, matured when he learned to accept stopping mid process and walking the long shifting bridge from next images to those left behind weeks ago.

He raised his head slowly and looked in a wide circle. He saw the crystalline creek waters, the brightest green grasses on its edge, the overhanging red rimrock, the Aspen and the Pines. He looked to be there, and he looked as he had practised for years, atop that Menagerie forest column, from the porch of "her" cave. He looked to find "her."

Quickened, he saw the form, a knee from behind a tree and what looked like a long tail. He got up and walked to the side to see what it might be.

A lanky white-haired man sat against the charred Ponderosa stump, by his side lay a dead cougar. He stroked the big cat and Clevalure felt his eyes.

"Ash," he said. "Jenny."

"My wife!?" he blurted as he rushed forward.

"She loves you."

"Who are you? Where is she?"

"I am still Titus." And he looked down at the form of the dead cougar.

There is a thousand year-old Persian miniature, a painting, of two people sitting together cross-legged in a courtyard. The background is in a flat, muted green with orange and cherry undertones. The geometric border would appear to be an equation of sorts. And the two people seated holding hands were Penny Frill and Benjamin Aboo. Climb into that painting and hear them, Penny says "Titus is not dead, he's visiting death. Enno is not alive, he is just visiting."

Yes, says Benjamin, and I hear Enno saying "I do this not because God commands it or requires or would be pleased for me to do it. I do this thing out of desperation, for our God is dying." And Enno blinks silently.

And the museum visitors who stand before the painting do not hear these words, they hear nothing, but they feel themselves well up in its presence, and they feel time overlap.

As he sat where Titus had sat, the large dead cat's head cradled in his lap, tears of release dripped from his chin, he watched the thin man walk away and he felt an image of his first sighting of Enno Duden, saw him as a ghost, saw him beside Titus as a ghost. Ash Clevalure had work to do. Purpose had hold of him, it would need to share him as he continued to weep, continued forever.

Curator at the Rijeka City Museum, 'magnificent' Kristina Pavec posed on the upper deck of "Galeb," the yacht of late Yugoslav leader Josip Broz Tito, at the port of Rijeka. Once a stage for geopolitical dealmaking and host to the 20th century's most glamorous stars, the now-dilapidated yacht is set for a new chapter as a museum. Croatia's northern port city of Rijeka, which bought the boat nearly a decade ago, plans to transform it into the star feature of its stint as European Capital of Culture.

The auction clerk reading the tickets and adding up Jackson's total:

  Lot 19, bucket of hammers, $20

  Lot 115, 3 rolls fencing, $15

  #116, pile of posts, $35

  #126, belting, $10

  # 17, tray of bits, $255

  #129, pallet of baling twine, $40

"That was apiece of luck." He said to Duden

  #133, roller, $110

  #141, eveners, $35

  #155, contraption, $20

"What's that for?" Enno asked. "It's a governor for an Oliver 88 tractor, to maintain belt rpms. Like for threshing. Got lucky, nobody else knew what it was." answered a beaming Jackson.

  #167, tractor, $1,500

  #121, tires, $120

  #156, cider press parts and pieces, $55

"And that?"

"That my boy is for fun. Yummy gonna smile big."

Duden looked down on his fist full of tickets. He wanted there to be another auction real soon.

The Norwegian Academy of Science and Letters announced it has awarded this year's Abel Prize — an award modeled on the Nobel Prizes — to Karen Uhlenbeck, an emeritus professor at the University of Texas at Austin. The award cites "the fundamental impact of her work on analysis, geometry and mathematical physics."

One of Dr. Uhlenbeck's advances in essence described the complex shapes of soap films not in a bubble bath but in abstract, high-dimensional curved spaces. In later work, she helped put a rigorous mathematical underpinning to techniques widely used by physicists in quantum field theory to describe fundamental interactions between particles and forces.

In the process, she helped pioneer a field known as geometric analysis, and she developed techniques now commonly used by many mathematicians.

As Dr. Fil walked outside he noticed dead birds, just a couple, then dozens. He looked up and saw dark clouds slapping each other until sparks flew. He heard a child crying. And he hurried. Purpose had him by the throat.

He is foul, that man who loves the sounds of his own thoughts. But close to ends, that is what he is stuck with. And truth is, often he does not love his own thoughts. He swings hard and wide and does damage to all the world - as does most every man and woman. That the world's life span would/could be determined by the pointless needs and lousy demands of unstable egos is a cosmic joke of terabyte proportions. We are, each of us, as insignificant molecules - yet hold the ultimate cancer in our empty hearts. We demand plot, goals, trajectories, purpose, diets, legacies, enemies, communities, curable diseases, and property rights. We demand explainable outcome. We want endings to feel like endings. We want our arbitrary community to agree on a set of lies that would result in an acceptable outline, an acceptable identity.

"I do not believe in God and I am not an atheist."
    - Albert Camus

Chapter Sixteen

brown rock

"...don't cry, you idiot! Live or die, but don't poison everything."
- Saul Bellow, Herzog

"But I welcome the darkness where the two eyes of that soft
panther glow. The darkness is my cultural broth. The enchant-
ed darkness. I go on speaking to you, risking disconnection:
I'm subterraneously unattainable because of what I know."
- Clarice Lispector, The Stream of Life

Religions do not create, they take and they impose.
Farms create, they seldom take and never impose.
If God needs a place to heal, best he should choose forest or farm.

D r. Fil found Benji sitting in the library courtyard with Penny. At first sight they seemed like sculptures, immobile, seated holding hands, heads bowed. Were they praying together? But that was interrupted when they felt his approach. Penny sensed, 'this one is alive and in a hurry.'

Fil was relieved when he learned that Ashwan had started the process of securing the waters. He tried to explain to Benji how his discoveries were gaining speed and that time to fix things was running out. Penny inserted, "All of nature is at the ready. Notice the sky, the healing is already underway. Perhaps not for mankind."

Dr. Fil hurried himself and fumbled around in his normally well-organized brain when he said, "but I fear this war between man and nature, I fear it."

"That war, my learned friend, will ever be between men and men, nature is but collateral damage. She does not fight back. She shakes, shudders, boils, swells, waters and blows to find her balance. Man's battles to hold on to his obscene false notion of entitlement throw her off balance, yes. But he is pond scum to her. Man is but a small part of nature, she grants him a survivable environment through residual chemistry. She would laugh if she had a sense of humor. She does not do battle with man, she is but a landlord." Benji looked at Dr. Fil while he spoke.

"I see it differently. I see the battle between nature and man. I would agree we cannot win. What I am saying is that unless human kind backs away, concedes without reservation, the environment you speak of will make our continued existence impossible." Fil wiped his face repeatedly as he spoke.

"Our paths parallel, many parallels, we must disagree in perfect unison." offered Aboo.

"Can you set up a meeting? I need to talk with Ogdensburg, Shoulders, Parks, Fork, and Maltesta. Together."

"How do you know of these men?" Asked Benji.

And Dr. Fil surprised him with the answer.

Penny felt like a songbird waiting patiently on a branch for the Redwinged blackbirds to finish gorging themselves at the feeder.

Back at his lab, Dr. Fil, conflicted by arguments he was having with himself, left aside the nature versus man question and returned to his discovery of the spinning subterranean waters. He remembered something and found it in his reference books and read...

An aeolipile (or aeolipyle, or eolipile), also known as a Hero's engine, is a simple bladeless radial steam turbine which spins when the central water container is heated. Torque is produced by steam jets exiting the turbine, much like a tip jet or rocket engine. In the 1st century AD, Hero of Alexandria described the device in Roman Egypt, and many sources give him the credit for its invention.

The aeolipile Hero described is considered to be the first recorded steam engine or reaction steam turbine.

*Pre-dating Hero's writings, a device called an aeolipile was described in the 1st century BC by Vitruvius in his treatise **De architectura**; however, it is unclear if it is the same device or a predecessor, as he does not mention rotating parts.*

The aeolipile consists of a vessel, usually a "simple" solid of revolution, such as a sphere or a cylinder, arranged to rotate on its axis, having oppositely bent or curved nozzles projecting from it (tipjets). When the vessel is pressurized with steam, steam is expelled through the nozzles, which generates thrust due to the rocket principle as a consequence of the 2nd and 3rd of Newton's laws of motion. When the nozzles, pointing in different directions, produce forces along different lines of action perpendicular to the axis of the bearings, the thrusts combine to result in a rotational moment (mechanical couple), or torque, causing the vessel to spin about its axis. Aerodynamic drag and frictional forces in the bearings build up quickly with increasing rotational speed (rpm) and consume the accelerating torque, eventually cancelling it and achieving a steady state speed.

Typically the water is heated in a simple boiler which forms part of a stand for the rotating vessel. Where this is the

*case, the boiler is connected to the rotating chamber by a pair
of pipes that also serve as the pivots for the chamber. Alterna-
tively the rotating chamber may itself serve as the boiler, and
this arrangement greatly simplifies the pivot/bearing arrangements, as they then do not need to pass steam. (- Wikipedia)*

The good Dr. could see through to the applicable variant. He could imagine the geometry and architecture that would translate his discovery to an entire other sustainable set of energy sources and engine constructs. He wrote down *vetruvious infundibulum*. And added 'are we to continue?'

In Portland, Oregon, 125 chickens were discovered one morning running loose in the posh Pearl district. Animal control officers, children under the age of eleven and two old women scurried around for five days trying to capture the errant birds. Occasionally, usually on a piece of grass or a pile of rags, an egg or two would be found. The local radio reported that police required the eggs be surrendered immediately as evidence. There were chat shows speculating on whether or not the poultry had been set loose at night as an act of terrorism. Several overnight experts were spewing a theory that the birds were carriers of a particularly virulent variety of influenza. One homeless drunk had been interviewed saying that, while he was being held for disorderly conduct, he watched officers trying to cook the eggs in the station microwave. He heard the janitor lady yelling at the officers about having to clean burnt yoke off the insides of the oven.

On the fifth day, police raided a homeless cluster, under a highway overpass. The inhabitants were arrested for cooking chickens over an open burn barrel. The barrel-roasted chicken meat never showed up at the precinct.

That night two officers in the Burnside district were attacked by what they thought were dogs ripping apart chickens. In the emergency room it was determined that the scratches had been caused by coyotes.

The radio station got a late night caller who claimed to be an area farmer who had turned the chickens loose in the city as his part to help the homeless. He said, "I figured they could cook a few, and keep the others to raise eggs. I just wanted to do my part. I am so sorry that it caused trouble. I won't do it again." No one ever discovered who he was.

Two days later a small flock of sheep were found wandering the park strip in front of the Art Museum.

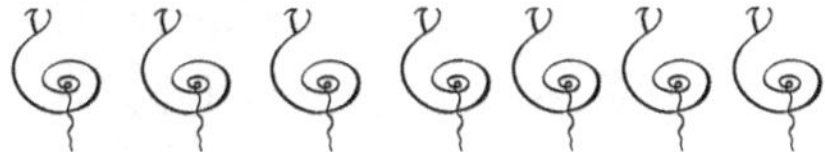

Vic Nib had his dog bite wounds dressed at the emergency room and refused to tell about the attack. He then went to a farm and ranch supply store and purchased an automatic rifle, ammunition and a machete. Next he went to the convenience market to get some cigarettes, carrying an unconsummated load of rage in his gut.

Finding the parking lot full of motorcycles and funny old men in leather vests and ponytails, he drove into several bikes and got out of truck. He was met by a circle of angry bikers. "You do not want to mess with me!" he yelled, only to have a tall beautiful Ulrika, length of hose in one hand and a cigarette lighter in the other, saying, "Oh, but I do."

With lightning speed she hit him across both ears, both shoulders and both knees with the hose. Then, as he stumbled backward, she walked up to him and lit his sleeve and pant leg on fire. As he swatted at himself to put out the fire, she kicked and hit him again and again.

A police car arrived and the bikers all leaned against its door to prevent it from opening. The officer called for back up. Meanwhile Tronk was coming out of the store to hear Vic shouting. Gasping and leaning forward with a terrible smile, Vic hissed and whispered to Ulrika, "I think I love you."

It would be the last words he would ever utter. Tronk took out his razor knife, slammed Vic's head against the store wall and cut off his tongue and threw it in the dumpster.

Then with a signal from Ulrika, a chain was used to secure the cop car door, and all of the bikers rode off with Vic's wallet, new rifle and machete.

When the second patrol car arrived, Vic Nib was recognized from cow-cop Fred English's posted description. The report named him as a danger-ous mercenary. Nib was taken to the hospital under guard and charged with a bucket load of felonies.

Under sedation, strapped to the emergency room bed, police at the guard, Vic Nib thought 'now this really hurts.' A lifetime in search of the worst pain finally satisfied, he tried to smile but it was impossible.

The cops, all locals, carefully followed the bikers from a distance until they were certain they had left town. Then they went back to the SnoCap drive-in to finish their burgers.

The article in the local newspaper did no justice to the occurrence.

Back at Snake Flats, Drummer found Uncle Nimbo and his dogs gazing at a short stretch of freshly stacked stone wall. A string line extended from the beginning corner on down seventy-five feet. The short section of wall was a thing of beauty. An assortment of lava stone, no two alike, had been 'organized' to perfect interlock in a flat face and top, three foot tall and eighteen inches wide, it tapered where it waited for the next additions.

"Did Enno do this?" asked Jon

"We watched as he and Benji set the string and talked about it. But the stacking was all Benji."

Jon Drummer picked up a loose stone and turned it in his hands while looking at the trailing end of the wall. Then he went over and carefully inserted it until it disappeared in its fit.

"Nice." offered Nimbo.

It was easy to find stones, this was a stoney land. Jon picked up two loose ones and returned to the wall to find where they should fit. After he did that he turned around to meet Nimbo who handed him two more. They worked this way for a few minutes then found Mikey, the big dog, coming their way with a stone in his mouth.

Jon was wearing an enormous grin. "Now this," he thought, " is how God meant us to apply percussion, in four dimensions."

Fred English was on the tail of Jeb the hit man, who was getting mighty frustrated to see the round faced cowboy/man pop up everywhere he went. It was kind of funny in the beginning; Fred thinking "this is a damn sight easier than keeping track of a cattle rustler." And Jeb chuckling "this buzzard has no idea who he's tangling with." But Jeb was anxious to get on with his contracts, Duden and the Ruskie.

Leo Vitch, learning through the grapevine that a hit had been put out on him, devoted all of his attention to identifying who the assassin might be. Small town, lots of tourists, but still and all it shouldn't be too hard to pick out a professional, just watch for oil slicks.

He spied an old local, slouch hat and stained Carhartt coat. Something about his pouring peering eyes, but naw he couldn't be the guy. Obvious, he was one of those holdout forest farmers who came to town once a month for supplies and worried after all the new people. Nope, Leo was looking for someone in city garb who looked like a Serbian butcher with Irish eyes.

Sloop saw English and could tell the retired cowcop was on a case. Following Fred's eyes he saw Jeb and clenched his teeth, reaching into his deep coat pocket. In that instant, Leo followed Sloop's gaze and finally found his mark, Jeb. Fred English caught sight of Sloop and then this new character Leo. They were triangulated, and rotating, each sought cover and clues as to their next action. What triggered the next sequence was when Jeb finally laid eyes on Leonid Runaroundavitch aka Leo Vitch aka The Sausage Doctor of Tralinka aka Lieutenant Bloody Sockets aka Meiko the Merciless aka... Six shots, four men down, each with a bullet in the left shoulder. One of them was being beaten by a silver-haired woman with a stout umbrella.

"You filthy man, you aren't going to shoot anymore in this town." And her last blow knocked Fred English senseless. All around people were ducking, running and screaming, all but the old woman who was cursing a blue streak.

Olivia (not her real name) had come to Mascara late one night in the Fall. A friend of hers had a house-sitting gig at Black Crater Ranch and when the press had gotten wind of the President having a 'kept' woman, a curried mistress, a toasty tart, a bling-worthy fling, living in a fancy two bedroom caddy shack on his small-ball course in Jersey, Olivia (not her real name) flew the coop and snuck into the house-sitter's digs. It was there that third party Filson Wool came under the influence of Olivia (not her real name). It took a short intense, getting to know one another, time before Filson demanded they do something about her name. He was fine with Olivia, but (not her real name) was dumb for a last name. He was intrigued with the fact that she was hiding from someone or something, it suited his twisted sense of fate.

She loved the fact that he was all screwed up with crankiness and that he generally hated everyone and everything. Suited her need for a challenge and cover.

She was tired of being picked out, picked up and offered money for her companionship. The last four years, every weekend with the small-brain President, she had wondered if she would ever rediscover mystery.

Then here was her chance meeting with a big dufus crank, Filson Wool. Someone who wanted nothing from her, and most likely couldn't afford her anyway. Someone who lent a crude luster to notions of trailer trash, some-one who put dijon mustard on his microwave burritos and ate them with Perrier water and Norwegian wafers. Someone who absolutely loved arguing with people just as sorry-assed as he was. Her chance meeting.

That was it! She changed her name, right then and there, from Olivia (not her real name) to Olivia Perchance. And then she told him.

"Up until now, I think it was safe to say you were destined to never amount to anything. But that is about to change. From here on its you and me bub. Or should I say, I'm taking you on as an extension of me. Do what I say when I say it and we'll be better every day. Otherwise you're fish food." She said this while wiping her hands together, and she avoided his dull eyes.

He said, "huh?"

They had both left out any thought of romance or intimacy.

"Where are we living?" she asked.

He imagined himself in hospital wishing he were home and able to sit before meals of his choosing, inside a room temperature of his preference, with lighting that gave musing options.

It **had** been a splendid life; good health, everyone a friend, but then almost without warning, he was alone.

His became a brittle itch so he shut down and, at 105, he died.

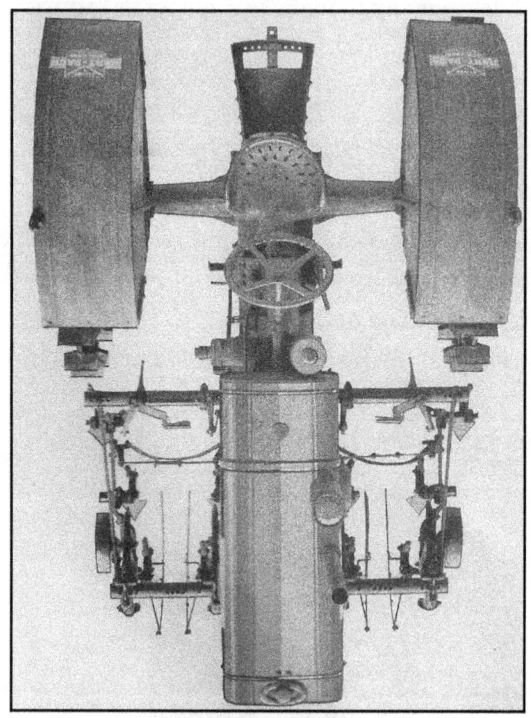

Some things you can't find out; but you will never know you can't by guessing and supposing; no, you have to be patient and go on experimenting until you find out that you can't find out. And it is delightful to have it that way, it makes the world so interesting. If there wasn't anything to find out, it would be dull. Even trying to find out and not finding out is just as interesting as trying to find out and finding out, and I don't know but more so.

- Eve's Diary, Mark Twain

Chapter Seventeen

grey rock

Ouaricon, Ouragon, Oregon, a mythical land of high winds,
herb gardens, dusty sage, a ship-eating coastline and forests
thick as hair. A place in waiting.

"it behoves you to develop a sagacious flair for sniffing and smelling out
and appreciating such fair and fatted books, to be swiff: in pursuit and
bold in the attack, and then, by careful reading and frequent medita-
tion, to crack open the bone and seek out the substantificial marrow
— that is to say, what I mean by such Pythagorean symbols — sure in
the hope that you will be made witty and wise by that reading; for
you will discover therein a very different savour and a more hidden
instruction which will reveal to you the highest hidden truths and the
most awesome mysteries"
*   - François Rabelais, Gargantua and Pantagruel*

The big black mule team were as different as salt and pepper, not to look at but in their posture, manner, intelligence, and appetites. They had been Titus' favorites. He named them. Always on the left, "Rubber." Because he rubbed, on the tongue, on his team mate, on anything he was tied to. He rubbed.

Always on the right, always solemn, thoughtful, patient and strong, was the mule Titus titled "Desmond Tutu." He felt it a title rather than a name. Titus Ibid had once met the peacemaking archbishop and carried inside himself the soul imprint of the man.

But names slip and slide. When in the barn alone, or sometimes working with Jackson, and referencing the span he would call them "Toot'n'Rubber" and the children loved to call them "Tootinrub." But it was Titus' want, when appropriate, to call them by their full names and never together. It was always Desmond Tutu, then, and with no 'and,' it was Rubber. That is how Enno was introduced to them.

They were hitched to a six foot single disc harrow. Titus talking to the boy for one of the first times, stood at their head checking the bridles. Standing near, Rubber rubbed his head on Enno while Desmond, head held high, looked down out of sideways eyes. Titus flicked Rubber's chin with a soft finger touch and said 'quit.'

"That's Rubber."

Hat in hand he would then offer, "Here. Please allow me to introduce you to Desmond Tutu."

"Beautiful team," Enno observed.

"I understand your compliment but I really don't consider them a team. I asked them to work side by side and they frequently indulge me. But though they each, individually, sense and respond to me, I have seldom felt or seen them respond to each other. Rubber is all about Rubber. Desmond Tutu is all about fairness and equity. When I work them I understand that I am working two single mules, I call it my side-by-side hitch."

And now, Jackson was to gift Titus' span of mules to Enno.

"This is hard," thought Duden, "it means that I must see him gone. I do not see. I feel, sometimes with understanding, many times without. But I do not see him gone."

"There's something wrong with him." They were in the lounge, back behind the restaurant, drinking beers and complaining, arguing sports and talking lofty stuff.

"I wouldn't have noticed him at all except for my daughter. She says, 'Dad, he's like some sort of spiritual savant or something. I mean he's like not really there except he does things that seem to suggest he's very wise, or knows something deep that's consuming him. One minute it's like he's an idiot, next he's a smart old judge.'

"She's tried to make conversation with him because he seemed so bright - different from the other young men around town."

"Says, 'And Dad I haven't seen him drool once.' and with my daughter's looks you know she gets more than her share of drool."

"Yeah," eyes rolling, belly chuckling "you can say that again."

"Come's in the hardware store where she works and, as she put it, walks to what he wants like he's taking the last walk on death row. Dragging his feet, but bolt upright, with innocence, looking around as though it's his last time on earth."

"Pretty extreme and weird yeah. Peggy is sensible, she wouldn't say it unless she felt it."

"Well one day I'm visiting her in the store and she whispers 'there he is' as this striking young man walks straight and slow, kinda fits her description."

"I was talking to Gunny down at the parts house and he knows him. Says he's retarded, also says the kid drives him nuts because he acts so uppity. Says the kid is trying to be a farmer."

"Bothers you a lot, don't it?"

"Well, yes, I wanna know. Who is he? What the hell is he trying to prove?"

⁂

He drives a new Lexus, every year. Lives side by side, not together, but *with* the woman he's married to. She has no working identity except to be intensely concerned for the welfare of anyone who wants to be her. He heads up a regional office for a global investment firm. They live in the gated golf course community just outside of Mascara. Most days he shows no emotion, every step all manners in tight control. He would prefer to wear tailored suits but this rural resort town, and thereby most of his clients, golf, fly-fish, jog, kayak, and drink craft beers. The best he's been able to do to fit in, and still not disrupt his cultivated manner, is to go three times a year with 'the' wife to Paris and get fitted in *Provencal* casual.

They are successful. They are empty. So much so that for the purposes of this narrative we shall see if they can go nameless - for, you see, for them and so many like them their names do not matter. Yet, on occasion they by default seem to be frequently chosen to decide the most delicate and earth-shattering of questions. So you might say superiority has its rewards, though we won't.

ocℐ∞cℐ∞cℐ∞cℐ∞cℐ

Enno was telling Benji about the mules and wondering aloud how to make his place suitable. Fences, a shed or stable, and a place where he could stay out there with them, actually living on that land, all such things seemed hurried and daunting in the moment. Driving in to the acreage, and talking, they were distracted, so it was Resumé who directed their attention with his yelping. Then they saw them. Elk spread wide, across the dirt road, on either side, out ahead of them and even behind. Had to be hundreds. Slowly walking, almost like camels, with a rolling, hips-in-charge, front-out-of-sync-with-back gait. They didn't seem to mind the old pickup truck. Then they formed a unified carpet and slowly flowed to the left, disappearing in clusters as they entered the steam fog belt.

Enno, Resumé and Benji sat quietly in the truck and watched. Suddenly the last of the elk bolted and they herd the squeals of young ones. There! In front of them, two large dark canines, wolves. Just as quickly, the predators vanished.

Not apostles, perhaps disciples, more like followers, they began as a group of protectors who were keenest about facilitating. They in unison saw the dream and wished to help manifest it. They in unison saw the threats and worked to neutralize them. They in unison saw him as incomplete but no less energizing and exemplary, and so were more accepting of themselves. They saw his light, felt it fragile, felt it warming, felt it vital and then circled round it, backs turned in to shield it from the winds. First they thought for sustainability they might add fuel to its flame but Benjamin Aboo showed them how that might burn it out, how it was that this flame was not a flame but a long, low, whistling, centuries old spark that actually did not give off light nor warmth but instead meant life. The way Benji showed them to protect Enno's light was to hold life sacred and mysterious.

Titus Ibid felt himself furious. He agreed with Jackson, Enno should have the mules. But he was as mad as a ghost dare be because he felt so clearly that he had not been done. Yes, he had chosen to trigger the explosion, because all of them had to be saved, and he was the only one to do it. He and Glass. He had done the right thing. But Titus Ibid had been a virile and sharpened old man with a swollen list of to-dos. It had not been his time, he kept insisting. Old is one thing. Done is quite another.

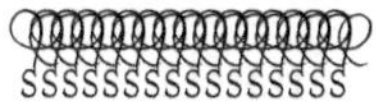

Car radio: *A man suspected of being a rhino poacher was killed last week by an elephant and his remains devoured by a pride of lions at a South African park, officials said.*

Rangers at Kruger National Park and other searchers found only a human skull and a pair of pants, the park said.

"No. Houston Tuttle is no problem. He's a fool. And the best way to neutralize a fool is to find him a job way above his station. As a silly-as-shit, figure-head president the media will dissect him over and over again. Don't worry about Tuttle.

"What we're worried about are these mumblings across social media that there is some sort of second coming out in the woods somewhere. We've got heat readings in South Sudan, Oregon, plus the perennial mumblings out of Arkansas and Tibet. Someone pretending to be prophet, or another Jesus, riling up the poor and working classes, that is purely bad news for us. Corporate governance cannot survive a second coming. Whatever it costs we have to drive this to ground. Set up a meeting with Colonel South. We need Grey-Water. The RBMC (Riksdag-Bundesburg-Murdoch committee) needs to be notified."

10 PLO PLO PLO PLO P

<This book is insane. Nobody is going to believe this junk.

It was never meant to be believable. It was to be provocative entertainment, it was meant to be thin sweet soup with floating clusters of thought dumplings.

Well, it ain't. It's more like Edward Scissorhands alone in a dark closet with a roll of wet toilet paper.

You know I can find other narrators.

I'll say it again, this book is insane...>

Shut up.

/\/\/\/\/\/\/\/

Shirley said, "We have to hurry now."

Sig responded, "I know."

"No, I mean we really have to hurry now. I feel something slipping away from us."

hehehehehehe

When his wife had left him, she took all the electronic devices. It had been ten years ago. Not so long. When he had moved into this double wide, in the mobile home park, it had been clean and empty. His job at the Habitat Store had begun with him feeling like he was a pin-setter at that very old bowling alley in Cozad, Nebraska, where he had grown up fifty years ago. He could never turn his back. Had to always watch the bowlers as he did his job. The recycling place was like that. Seems people come in with their junk and a "I'm better than this place" attitude. They just drop or fling stuff like it's the dump and they wanted out of there quick.

Eventually though, the job had allowed him to gather furnishings for little or no money. Never had to worry about decorating elements, everything that came through the rehab store just seemed to fit the trailer house perfectly. It was at this job that he developed a passion for clocks, the electric, battery and windup kind with dials. Could never get enthused about the digital ones. After ten years he had thirteen clocks scattered around his small home. All of them working and all of them set to a different time. He never changed them for daylight savings, he never adjusted them when they ran fast or slow. He got a chuckle out of the fact that the time they told meant little or nothing. Made him feel like he had time under control for the simple reason that it adhered to no standard.

But still and all, with no tv or computer or telephone, he found he had too much time on his hands so he took a night shift as a dishwasher at a Mascara restaurant on main street. This worked out well, because he could eat on the cheap. And he enjoyed, on slow nights, visiting with the regulars who would sit at the restaurant counter, eat, talk politics, flirt with the waitresses, and drink coffee. They became his chums.

His chums, and the waitresses, knew him by his last name, Cochran. Eventually, because of his time-sensitive apron patterns, they gave him the nickname of Smudge and it stuck.

Smudge Cochran was thrilled the day a regular customer had asked him if he wanted a car. Up till then he had walked or biked to his two jobs. But now, here was an old beat-up Datsun pickup he could have for just going to get it. Smudge Cochran's life was in balance, far from complete, but balanced. At least up until the awakening. One evening while busing tables he had overheard a conversation that shook his simplicity to its core.

"I don't agree with you. Not at all. As I see it if a man wants to farm that's a good thing. Hell, in this day and age if a man wants to do anything that might be remotely honorable that's a good thing." The man talking was not a regular to the cafe, he 'felt' like he was passing through, visiting town. Maybe a salesman? The other guy Smudge knew. It was Gunny from the parts house. He thought of him as 'Grumblin' Gunny.'

"Farming? Next thing you're gonna tell me that we get food from farming. Man, don't you read or get any news? Farming died around World War II. Now it's all Agribiz. Drones signalling driverless tractors over humongous fields where everything is fake; fake seeds, fake fertilizers, fake poisons."

Cochran's brain started to slide, then it went over the edge into a tailspin. No one had told him. Farming dead? No one had told him...

He was sitting on the tire in his Uncle's Nebraska farm shop watching adjustments being made to a seed drill. Uncle Thaddeus was explaining to him the honor in farming. Young Cochran wasn't listening, but he was hearing. And now all these decades later he was so terribly sad, missing Uncle Thad, missing the sight and feel of the big man's hands etched with oil-filled work lines, missing the smell of dried chicken manure, missing farming.

"Well, I say if this young man you're talking about is trying to build for himself an old fashioned general farm, it's one way the world has a chance."

Gunny Sax choked on his fry, trying to express his complete disgust.

Chapter Eighteen

gravel

"The most beautiful thing we can experience, is the mysterious. It is the source of all true art and all science."
- Albert Einstein

"If every day a man takes orders in silence from an incompetent superior, if every day he solemnly performs ritual acts which he privately finds ridiculous, if he unhesitatingly gives answers to questionnaires which are contrary to his real opinions and is prepared to deny his own self in public, if he sees no difficulty in feigning sympathy or even affection where, in fact, he feels only indifference or aversion, it still does not mean that he has entirely lost the use of one of the basic human senses, namely, the sense of humiliation."
—Vaclav Havel, letter to Dr. Gustav Husák, 1975

Plumley & Recto formed an Americana band as two gay men writing fling-modern feminist anthems utilizing dobros, Scottish clarinets and the percussive sounds of kitchen utensils. Plumley came up with the idea. Neither Plumley nor Recto (real name Charley) were gay, and they had no sympathies for feminist anthems. They did own two reasonably useful falsettos. And they knew the time was over-ripe to be on the carpet-side of history.

They naturally evolved into internet vultures looking for easy money and dramatic attention. So when their music started to drag, pun so obvious, they latched on to blogging about religious fanaticism and culture wars, their over-educated perspective being neither right nor left but a nasty downward slant. It was particularly wild for them when they got hold of crazy little rumors of government and corporate board rooms actively concerned about religious fervor and trying to hunt down messiah types.

Plumley had devised an internet search process which pulled in huge amounts of similar hits in triangulated pulse which sought not to find *some* thing but rather to find commonality of what nervous networkers were looking for, or excited to hear of. Patterns started to emerge with particular intensity. Doing daily overlays they came up with an energy they nicknamed "The Engine of Quanchee." They understood that the address of nervousness held great power.

It was a zealous Recto who, in deep historical digging, discovered the century's old humming magnetic force of *Gendun Drub*. And then, with great disdain for the truth of the biblical Enoch, Recto guaranteed his place in the history of slime by coming up with a nasty descendant he fully imagined as Enoch Drub.

In all the searches, Plumley only came up with the name Enno Duden twice in relation to an emailed corporate memo. But the name intrigued him. The sound of it, eeeno dooodin. It could have been the name of a shy hero from some children's book. Together Plumley and Recto found themselves whispering in harmony *Enno Duden - Enoch Drub - Enno Duden - Enoch Drub* and they lit up with bitcoin smiles. Only thing was, they could find absolutely NO record on-line of who this Enno was, where he lived, if he lived, his history, pictures, nothing. And that, after a moment's pause, secured it. Here they had the eye of the tiger. Now to flesh out their imaginary Enoch Drub.

<I'm out of here. I can no longer be a party to this.>

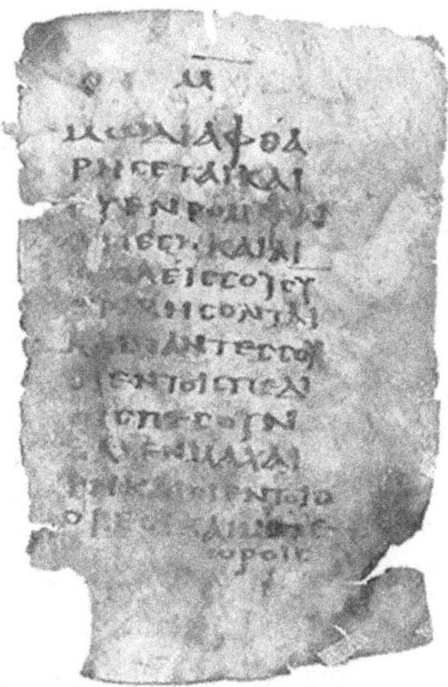

Aren't you getting a little tired of yourself. If what you say is true then leave, just leave. The story will take care of itself.

<   .>

There were those middle years when fear would wash over him as he worried that promises made several lives ago would not be fulfilled on this trek.

"Hey, That was way too easy, I found you. I know you. You're Jon Drummer."

'Here it comes,' thought Drummer. It was sweet while it lasted. But, dammit, I'm not ready to go back to the band thing just yet. He was sitting at the counter eating his fried trout and eggs and watching Smudge Cochran wipe the counter when this guy walks up and slides in next to him.

"My name is D.E.B. but you can call me Dude. Please don't freak out but what I'm going to tell you is gonna sound super weird. Last things first, Duden is what he says he is. He's just a really good, strong kid with no baggage and he really does want a farm of his own. Nothing more. And you are what you think you are."

Smudge over-heard this and slowed his counter wipe to a pressure swipe. Dude looked up.

"I know you heard that, Cochran, and I'm telling you, you need to take a break and stand right here and take this in because it's your life too.

"Only way for me to do this is to get it out quick, in a blast, before it gets erased or worse. You guys are in a story, a novel, that this psycho is writing. It's not a psycho novel, but he is. And I say he is because up until a paragraph or so ago I was his narrator. I've been watching this story come out of him in bursts. And I'm seeing a pattern. First much of the line is coming through him, and now I realize, sitting here talking to you, that it's real or as real as it can be. But then there is other stuff that I can feel him making up, strange stuff that is pushing the story towards terrible terrible endings."

Jon is all in, listening intently, something is reminding him in a narrow fixed way of that man who sent him to Cultus Lake with the fishing gear.

Smudge is hanging on to the conversation by the barest thread, that farming thing.

"I know you both, I was a voice telling readers about you. I know every element that made it to the page up until him telling me the story would take care of itself. That can't happen. So I quit him and found you."

"Please hear this! The story is not about Duden! It's about the Brown Dwarves, both of them. It's only a guess but I figure the author is getting messages about this story-line and his nasty, depressed, and paranoid self won't just write that down, he's got to keep inserting his own stuff to direct the story towards his own vision of apocalypse. That's wrong! And the only thing I can think to do is to tell each sane character in this story the truth.

Just then, as Jon and Smudge were watching and listening, Dude disappeared, just went poof.

Next thing Dude knew he was walking on air with Titus, trying to tell him everything. But no words came out. Didn't matter though because Titus knew what Dude was trying to say. No sound but Titus said, you're my ticket back. So we learn that *the Dude Abrides.*

Lloyd had brought with him his three old leather medical supply and tool satchels full of his brushes, and assorted art supplies. He brought no clothes, no toiletries. He knew that Cash Apercu would see that these were waiting for him at his lodging. Cash had arranged for Lloyd to stay in a Greenwich artist loft that had once belonged to Cy Twombly. At the loft were the Levi and Carhartt clothing items Lloyd needed, plus two people who were to be his studio assistants, drivers, and body guards. These two spoke only when spoken too and never offered opinions. Lloyd did not want them there but found, after a little while, that they would save him quite a bit of time and worry. By his own insistence, he would need to complete his entry into the competition in a short time period. He was already worried enough about the ranch to make his first phone call to Jackson to check in. While he's talking on the phone, Cash arrives with a visitor, Anatoly Grim, who is immediately subjected to a full pat down by the studio assistants.

Grim is grim, looking like an early-retirement mixed-martial arts fighter in golf attire, complete with platinum bracelets and necklace.

Lloyd finishes his call and steps forward. Hand outstretched Grim says, "It is an honor and a dream of mine to meet you. I have three of your pieces. You are a great man. You know your work will be worth many millions when you are gone."

"Well then my life will have been a failure by at least one standard. Steinbeck said we should live so that no one rejoices when we die."

Laughing, Anatoly offers, "Please my new friend, understand my generosity when I say to you, I want you to live a long time and make many more sculptures which I will have to work very hard to afford. But, yes, I hope to outlive you and bask in the rapidly ascending value of Lloyd Shoulders. Oh, and yes, I know, deep down, many people will rejoice when I am gone."

Apercu stops the conversation: "This is why I cannot take you out in public. Now you have met him. I insist you stay away. Or, my oily adversary, I will give those minions their time of rejoicing."

"See, that is what I mean. Shoulders, you and I are in the best of hands, Apercu is the sentimental butcher we require."

Ecuador's Pululahua Geobotanical Reserve is within the only inhabited cultivated volcano, with corn, sugar cane, beans and camote - a rare potato variety.

The Moche culture of ancient Peru is 5000 years old, equating with Egypt and Mesopotamia.

Dr. Fil received a bulletin about the changes at the Tern Lake Thermal area at Yellowstone. He got in touch with two volcanologists via a conference call and gave them a hypothetical.

"The earthquake has caused shifts which have resulted in subterranean waters changing direction and merging, in part, with a new geothermal feature which is pushing the mixed temperature waters to the surface. This is all occurring in the foothills of mountains which are inactive volcanos. What metrics might be employed to determine the instability of the area? Could any of this be pointing to eminent eruptions?"

A piece of a *Science Day* radio interview was overheard while driving his Prius to pick up his girlfriend at her yoga class. They had a firm reservation for an exclusive LA restaurant for lunch, so he had to shut the radio off before hearing the conclusion:

"...don't take it seriously, I think that's because the explanations come with a thousand variables and all of the scientific jargon is laced with political defense. The biological world of the earth is convulsing and revolting all at the same time. Let's think about it this way: earth has a compounded infection on its surface, think of it as lice and rashes mixed with a terrible fever. So the earth shakes to scratch, storms to weep, freezes, blows nasty winds to cool itself and clear the air. The results are totally unpredictable and unprecedented patterns of weather; storms, heat waves, earthquakes, volcanic eruptions, floods, and poisonings. Follow this litany with disease and famine as agriculture entirely fails. The mass of humanity will suffer, thousands even millions will die. Migrations will increase. And the rich and entitled will migrate as well. The countries which will offer a chance at quality of life will be Canada, Scandinavia, and Poland. So one percent of the world's population will go there. The rest will..."

He found himself asking her, over their wine spritzers, "What do you think of Finland?"

At the trial of Besom Cellar these things were clear: he was a wealthy retired hog farmer who had gained access to the South Florida resort/compound of the past president and posed as a caddy with an ingenious system for cheating which he 'sold' to the president. He buddied up with him in this way and claimed to have arranged a secret golf match between the president and the previous president at a far remote private course. He had made arrangements with two disenchanted air force colonels who commandeered a C130 cargo jet to play the part as they took the president, and his gold plated golf cart, to the aforementioned secret golf tourney. As icing on the cake, the president was promised he would win and that the match would be televised on his favorite news media. High in the skies over Chiapas, Mexico, turbulence was faked and the president was strapped into a parachute. He was told his sacrifice was required for the country. They opened the hatch and pushed the gold plated golfcart from the plane. Next he was shown the pull cord on his chute, told when to pull it, and summarily pushed from the plane, never to be heard from again.

The two Air Force colonels were awarded national medals of bravery and alter egos and flown to asylum in Norway and Spain.

Besom Cellar chose to confess and stand trial so that the country could hear why he chose these actions. The presidents 22 million loyal followers wanted him hung. 200 million other Americans wanted him given the Congressional Medal of Freedom.

In the movie of the event, Besom Cellar was to be played by George Clooney, and the president was to be played by Bette Midler.

Besom Cellar would go on to become a national hero, frequently compared to George Washington and W. C. Fields. That president would forever be a small-type, ugly, footnote in the history books.

The first gospel of all monarchies should be Rebellion; the second should be Rebellion; and the third and all gospels, and the only gospel of any monarchy, should be Rebellion--against Church and State.

\- Mark Twain, Notebook

Chapter Nineteen

sandstone

In December 1837 Emerson shared a discovery with Thoreau.
The previous year (I had) "found a new musical instrument
which I call the ice-harp. A thin coat of ice covered a part of
the pond, but melted around the edge of the shore. I threw
a stone upon the ice which rebounded with a shrill sound,
and falling again and again, repeated the note with pleasing
modulation. I thought at first it was the 'peep, peep' of a bird
I had scared. I was so taken with the music that I threw down
my stick and spent twenty minutes in throwing stones single or
in handfuls on this crystal drum."

- borrowed from Lapham's Quarterly who,
before that, borrowed it from elsewhere

Jon Drummer took hold of the trembling Smudge's sleeve. "It's ok, really, it's ok."

"Do you know the man he's talking about, the farmer?" Jon nodded and said. "Do you have a car?"

"I've got a pickup."

"Perfect, sign out sick, you and I need to go right now."

Back there, during the Besom Cellar trial, there was mass confusion in D. C. which worsened when the new acting president (the oft missing vice president) abruptly resigned. That left the aging female speaker of the House of Representatives as next in line. She would ascend while the unusual and electric election of Houston Tuttle was in full throttle. Her first order of business, as short-term acting president, would be to sign deportation orders for the entire remaining white house staff and cabinet. They were to go to Brazil to assist that government's rush to deepest idiocy.

Corporate board rooms and the RBMC were in full disarray. The economy was searching for stability when Wells Cargo Bank declared bankruptcy and the New York Stock Exchange cloud files were hacked and scrambled by North Korea. The market, in free fall, dropped two thousand points, there was a run on banks, bitcoins waved bye bye, Google imploded, and Facebook officers were arrested in Switzerland to be tried in the world court for indecency, the first such case. To further the chaos, Thayer, the German pharmaceutical company and agribiz giant, announced it was divesting itself of agriculture.

Talk radio had theorized that Mexican drug lords were holding *him*, so Colonel South and the mercenary force known as Grey Water were pulled off the hits on incipient messiahs. Instead they were sent, by the president's political party, to Mexico to wreck havoc on the Mexican cartels and find what happened to the now long lost former president. They were joined by a caravan of white nationalists, three taco trucks and fourteen news vans.

The border was a difficult crossing these days because so many former Hispanic migrants were now reversing and trying to get back into Mexico. The U. S. had become a wasteland of economic collapse, it seemed that mother Nature had particularly intense anger towards the States, which now reported enormous storms tripping over, and compounding, gigantic storms. So a re-coagulation of the original 'yuge' caravan of former migrants was being joined by smaller but more impolite crowds of mostly white suburban families desperate to find some sanity, stability and perhaps a job teaching English in Juarez. The Mexican authorities were clamping down, refusing to allow anyone into the motherland who wasn't already fluent in Spanish and the tactical subtleties of bribery. But the greatest fear was held by a percentage of Americans who were determined to flee the States before the coming of that most hideous of monsters, 'health care for all.' Insurance companies were advertising that when it finally happened it would add to this, the new greatest depression, by closing all for-profit hospitals, clinics, and ED labs. Doctors, nurses, and worst of all insurance salesmen and executives would have nowhere to turn. They predicted the coming of an age of witchcraft and voodoo medicine.

An invisible Dude Ironymous Binge was whispering into the ear of the visible Jim Nimbus Sparkle (Uncle Nimbo) who in turn swatted at the noise.

"What did you do?" asked Helga as she changed the bandage on her husband's shoulder. The bullet had grazed, cutting a burnt groove.

"I started with some idiotic scare tactics, trying to head off this business. Then, when I saw one of them taking a bead on Fred, I fired one pistol shot. Seconds later I was hit. It's starting again, honey, bad people have come to town to try to stop Enno." offered Sloop.

"Like before?"

"Seems to be the same energy. So far I count three but I think there's more. I took Fred to the emergency room. He's ok. He's been on their trail for awhile. He got a message from Monk and Shirley, they're on their way back. I've got a contact that's monitoring several fronts. Lloyd's in New York City. That has me worried as well. I'm going to meet with Jackson and see if he needs help taking care of Lloyd's ranch maintenance."

"You know that Sig will try to go on the offensive, he's probably already doing it. I'm worried this will blow up again."

Sloop nodded, "If he's with Shirley, that's a good thing. She has a way of bringing calm." Outside it started to rain hard.

"John, I feel it's time to forgive the boy. Something tells me we all need to work together now. This has gone from too-close-to-home to right here right now. But no more Pascal business. You know how I feel about that."

Back at the shooting site; Vic Nib and Leo Vitch had managed to crawl off before the police had arrived, each keeping an eye on the other. There was an invisible cord tying them together, keeping them just close enough for a final battle.

Storms increased and converged: Three back to back hurricanes approaching Florida and the Gulf of Mexico - A massive storm front curving down through British Columbia and into the northwest states - Fierce lightning storms setting off forest fires in tinder-dry Maine and New Hampshire. A cyclone in Oklahoma. Polar bears in Fairbanks. Large scale flooding in coastal Florida, Mobile, New Orleans, Houston, Los Angeles, Seattle, Portland and the San Francisco bay area. Heavy rainfall aggravating saturated soils. Four massive earth slides closing California's Highway One in three widely separated spots, stranding entire communities.

The long term weather forecast called for a nationwide calming stretch. The bickering, overwhelmed and immodest federal government had ceased to be an aid. State and local governments, in efforts to ameliorate the damages and prepare for more climate challenges, over-turned or usurped federal domains, implementing their own life-saving measures and ordinances. Polluting factories were nearly all shut down by Martial law. This was not lost on several rebellious sorts, who saw this as a de facto fracturing of the United States. Efforts for Oregon's secession ramped up.

Social networks and news entertainment broadcasting stirred the steaming pot of devastation to an ad sales bonanza. If you were plugged in, it was impossible not to feel yourself being thrown from one tragedy to a phone company commercial to another tragedy to an erectile dysfunction ad, all while fearing what would soon happen to you and yours.

Millions of people sat at home and watched internet-connected screens, absolving themselves of complicity and forming millions of different conclusions on what was happening. Half of them, because of the storms, fires and floods, lost power. When their screens went black and no one answered service calls, they sat watching their silent screen and waiting. Theirs was a deadly draining patience, waiting for 'life' to reboot.

Hundreds of thousands of people, long ago unplugged from the internet and electronics, or too poor to have ever 'enjoyed' same, continued on with their work days, answering local calls to help neighbors if there were fires, or floods or heavy wind damage. If you were to ask them years later, they would say, "I remember it was a really stormy year."

Enno Duden was not plugged in. Had never been. He and his band of farm building buddies continued with their odyssey, navigating an unusually wet season with what looked like casual concern. Enno put everything else out of his brain. He felt the swell of an oncoming wave and knew it might give the farm's construction a boost.

Benji knew something destructive was trying to get close. Strangers were asking questions, twice in town he had managed to turn Enno quickly when he spotted someone trying to take his picture with a phone. Much much later, when he would write the Book of Benjamin, there would be particular cautions laced into discussions about technology's threat, especially so-called smart phones. But for now, working with his new confidant Ash, he had come up with a simple code word that their inner circle could use if they wanted a new, unfamiliar, person to gain quick access. The word was *Nicodemus.*

Elsewhere, trying to add substance to stinking muck, Plumley and Recto had also come up with a codeword for their growing inner-circle, a way to save time and identify when someone had shown they knew that *the Engine of Quanchee* was a hoax and a joke. That the Engine of Quanchee was a corroding personality in gaseous form. Their code word was *amfivolía*, the Greek word for doubt.

It took Jon Drummer less than ten minutes to figure out that Smudge Cochran was exactly what you saw and nothing more. He thought,

'He's kinda like Duden in that way, but the difference is so wide. Duden insists he's just what he is - insists it by his nature and work. Smudge follows his own inadequacies like a sobbing child wanting an unpleasant day to be over.

He doesn't insist he's empty, he just is and it hurts. Duden reassures. Smudge worries. But they seem to be magnetized in similar ways?'

Jon felt the beat of these thoughts.

"Buddy, listen, if we get separated and anyone presses you, just remember this word, *Nicodemus*. It's a code word that will tell the right people you are OK. It will tell them you are also the right people. Can you do that?"

Smudge looked down, worried at first and then felt a smile crawl across his entire face. Did this mean he had just been asked to join, to belong? This acceptance business was powerful medicine. But now, it couldn't be true, not him. But why not? He knew why not and he threw that away.

Enno was quick to say to Nimbo. "Benji says what we saw were wolves."

"Thought as much, my dogs were particularly alert. I don't know if our little friend told you, but this breed has been raised for centuries in the Pyrenees specifically to stand up to wolves. They haven't been legal in this country. Had to sneak in the pups. Heard that was changing. Don't worry about the wolves. As long as my boys are here, the wolves won't surprise us. But, with your permission, should they come around I will shoot them. I know that it is illegal and I will take responsibility for my actions."

Enno said nothing, but he looked concerned.

Benji offered, "As with everything about good farming, we just need to be constantly mindful of balance. Don't worry Enno, wolves, like most predators, will always go for the lame, sick and weak first. Uncle Nimbo may not agree, but I fear human hunters far more."

"You have no argument there little fellow," said Nimbo with smaller eyes.

The gramatic the dramatic the whole cloth the hole cloth
Frugal fughal bugle beagle in the bottle deep in the bottle
bug eating snug bleating niggling noses up
we go thirdly why not shortly anyway you look at it
we go through the math

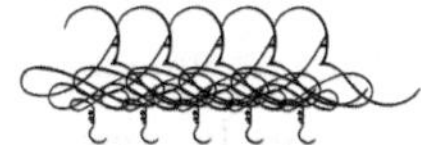

Lawyer Clevalure was visiting the salvage yard of Doobie Lexicon, spread across the width of a canyon just outside of Prineville. There were piles of tires stacked fencelike, for thirty feet, between rock jacks - these holding framework on which eight foot long sheets of used galvanized roofing tin stood upright for another thirty feet to another waiting rock jack - in this way creating a pattern, tires to tin to tires to tin. The gate standard was made of two huge gnarled Juniper trees topped in breech by two old extension ladders fastened together truss-like in glulam fashion. The gate to the yard was constructed of sections of 3 inch irrigation pipe which had seen service in a pasture, showing their trampled history with squashed and twisted forms. These pipes were wired to heavy duty hog panels. Rather than being hinged from the side, this heavy free-form, sculpted gate was hung from the laddered truss by rusting chains. Bottom center of the inside of the gate a 3/8" steel cable which ran in and up at forty-five degrees, to the top of an old drag line boom. Center of the gate hung two signs. One, on white enameled steel, in clear official black block lettering with thin red outline, read

YOU'VE GOT TO BE KIDDING.

Then underneath that sign was another bigger one, lettering by spray can, which read

YOUR SORRY ASS IS WHAT MY DOGS HAVE BEEN DREAMING OF BITING

Off to the side, a small sign said HONK AND WAIT.

Ash honked and waited. It didn't take too long. He heard a small motor and saw that the cable was being wound in to pull the gate back and up like a giant, scary flap. Out of the cab of the ancient dragline came a man in overalls holding short leashes on two snarling Rottweilers.

Ash asked, "Is it safe for me to come in?"

"Cool man, now that's an intelligent question. Who are you?"

"I believe I'm your attorney." With that the man unsnapped the two dogs, stamped his foot and watched as they slunk back into the dragline cab.

"You sure don't look like no Perry Mason. Come on in cat. Welcome to my digs. Drive your rig in here, make sure it's far enough inside that gate don't come down slap it in the butt."

At that exact moment a Sheriff's cruiser raced up behind Ash with its lights flashing, sliding to an angled stop. Ash saw this in his rear view mirror, when he returned back to the gate attendant he now saw a shotgun he hadn't noticed before. He quickly got out of his car. "Wait just a minute," he hollered as he turned to walk back to talk to the Sheriff crouched behind his rig, pistol in one hand, cell phone in the other. The sheriff yelled, "Drop the shotgun! You, old man, step aside!" Ash backed up slowly, got back in his car, and quickly gunned it into the salvage yard. The gate came crashing down before the Sheriff could get back in his car. Ash watched in the mirror as the Sheriff slowly backed away and drove off.

Back at Snake Flats a hard steady rain was adding to the slowly increasing flow from the two new artesian springs, one hot one cold. A decorative and designing erosion was well underway. The sandy glacial soils washed into side cracks and hollows as it was freed from ancient basins. Roping roots were exposed and used in new ways to help define the dynamic of the place.

The geography and elevations placed the river headwaters not in a hole or low spot but up on what would be exposed by the running waters as an enormous fifty foot round rock riddled with failed and successful pipelike tunnels. Off to one side of the rock there was a ten foot wide dish, four feet deep. Some of the pipe-like holes laced the sidewalls of the dish.

On the opposite top side of the rock, cold water bubbled straight up ten or twelve inches and cascaded over the entire surface. Down on the back side steam rose, as hot water ate at the rock base from a different source. That water, pressurized by Earth's ambitions, found its way into the piping holes. In this way the wide-rock side dish was being filled, from the top by cold water cascading over the rock's top face and, from the dish bottom, by hot water jetting into the mix.

Resumé may have been the first to discover this. Chewing his bubble gum, the old Australian Shepherd held his head sideways as he tried to understand what he was seeing.

Ash, got out of his car and greeted a smiling Doobie Lexicon.

"Far out. Like, you are some gutsy cat. Small but gutsy. We're gonna make a great team." Doobie introduced himself, "I call myself the Doge of do-over, the Sargent of Scrap, the Grand Wizard of Garbage. Only this..." he waved his arms wide, "this ain't no garbage. This is purest gold, this is the forgotten sacred promise of too-close-to-call dumpage and recycle. This is evidence that most humans don't have a clue what they got. Not even after it's gone."

Ash looked around, not at garbage but at chosen rescue. Every single thing seemed to be entire, complete, and, in an oily sort of way, beautiful. Over there were four 1950's vintage Ford pickups, used but appearing to be serviceable, parked in perfect lineup. Down one line of the farm fence, he saw old two cylinder John Deere tractors and Farmall M tractors. In a half circle around the back of the dragline were parked 1950's era Buick and Hudson cars. There were several old aluminum clad camp trailers of different makes, Oh yes, there were piles of tires, and partially disassembled vehicles and machinery, but all of that with what appeared to be a loving pattern of use, presentation and accessibility. And wandering slowly around everything were a handful of bearded goats. Ash's eyes continued to drift, everywhere he looked he saw treasures of other times. Then he noticed four different piles, neatly stacked and stickered, of used building materials.

He was surprised by a young voice, "Hey Doobs, where you want these?" It was a teenaged boy carrying four mid-sized steel wheels."

"Line them up neat like, over along the milking machine shed. Thanks Kiddo, you're the cat's meow." Turning to Ashwan, "Couldn't do this missionary work without my young friends." As the boy walked off the two Rottweilers followed wagging their tail stubs. In the distant corner Ash made out the forms of two more young people stacking something.

"I'm beginning to guess what you might need legal help with. But please, you tell me. I am prepared, and pleasantly so, to be surprised. Your 'digs' are indeed cool."

"Far out, I mean far out. Well, we got this new sheriff's deputy, transferred from Fresno or some such place, probably for killing cats or like that. I've been here for many years and the neighbors, especially the kids, love this place. It's like a genuine way station for stuff some of us don't ever want to see disappear. Dig it? Years ago I built the fence to keep the goats in, not to keep people out. Without the tin-eaters, that's what I call the goats - they don't really eat tin, I'd be over run with weeds. Then this deputy, Bart Prussian, shows up. He figures I'm hiding something in here. Doesn't like me or this place. That's cool, takes all kinds I always says. But after a while he's determined to see me run off. That's what those signs are about, just him. But he keeps pulling stunts like what you just saw. Daddy-o, I need to know what my rights are. I got this real creepy feeling something bad is going to happen and, dullsville, I will be a goner and my 'garden' will disappear."

"Doobie, this is something we can repair. I have several ideas. May I change the subject? Tell me about those piles of lumber."

"Those? They're barns. When the tech people moved in to build those giant juice boxes up on the flat they gave me those barns to remove. Paid my kids and we took them down, piece by piece, siding, posts, rafters, roofing tin - everything. Each pile is a barn. Couldn't let them just bulldoze and burn them.'

"How'd you like to see them put up and used again for a young man's farm?"

"Far out! Far, far, far out! Can we help?"

Dr. Fil's annotated, footnoted and enervating dream

It began with him as a child in a malt shop, banana split before him, the shop turned into a pawnshop which changed again into a car wash with a waiting room lined in chalkboards upon which he wrote mathematical equations until his brain made him stop and see this consequence: The nations of Brazil, India and China formed a secret cabal and, working through a maze of international corporations, arranged to purchase the state of Oregon from the U. S. Government for a price of eleven gazillion dollars inviting 5,000 well-positioned government thieves to dream of obscene wealth. The down payment was a cool 750 billion dollars. Brazil, India and China laughed their heads off knowing that a promise is just a promise and that the greedy gringos would have no recourse whatsoever when in ten short years the balloon payment was missed, then disregarded and subsequently forgotten. And the White House with its cabinet for life laughed their heads off muttering that they would have gladly sold that worthless troublesome hippy state for far less. Oregon became the world's first corporate commonwealth.

But that lasted only a year because the "board" of directors decided to sell off select assets and divide the spoils between the owner states. The red-woods were sold to Japan. Logging rights of the western Cascades were sold to South Korea, Fishing rights all along the coast and into the Columbia river were sold to Norway.

China claimed the northwestern corner which included metropolitan Portland, and ran from Astoria to Hood River. and south to the Russian enclave of Woodburn and formed the nation state of *Hinge*.

India laid claim to the central coastal portion running from the pacific rim to the coast and from Roseburg to Yamhill. This became the country to be known as *Shankar*.

Brazil claimed the remaining southwestern quadrant from the Pacific rim to the Pacific ocean, from the Siskiyous to Bandon and called it *Baziotes*.

The board room had allowed for the sale of the remaining two thirds of what is called Eastern Oregon to go into a consortium shielding the fact that the new owners were the primary investors representing Facebook, Apple, Google and Amazon. This landlocked central portion including Bend, Redmond, Prineville and Mascara was also divided into and incorporated as the high-tech state of *Decathlon*.

The residents of any unclaimed portions of the former state of Oregon were given notice of the changes and asked if they wished to apply to the consortium for their own territorial piece of the pie. The only two respondents were the attorney Ashwan Clevalure, representing a combine of all indigenous tribes on the one hand and a single individual on the other. The single individual was Enno Duden, who made an offer of one hundred dollars for a 200 thousand acre piece of high desert with a resident population of one. Sale was made and papers filed creating the sovereign state of *Sourse* and passing all rights and privileges for any surface water, mining claims, timber, wildlife and fisheries.

The Native peoples combine, calling themselves the First People, and acquired a patchwork quilt of reservation lands, they called their land the state of *Nowhere*.

All that remained, made up the eastern reaches from Lakeview to
Condon, from Millican to Jordan Valley. This became known as the desert
poor-lands of *Ridley.*

All in all the result was that the original investing three countries split a
profit of two trillion dollars with residuals stretching forward for all time.
And Oregon became seven countries individually known as;

Hinge,

Shankar,

Baziotes,

Decathlon,

Nowhere,

Ridley

    and Sourse.

China set a goal of increasing the population of *Hinge* ten fold, as did
India for *Shankar.* Brazil set its goal to log off all of *Baziotes* and plant
Poppy fields. *Decathlon* became a giant, gated, membership community
with zero emissions and genetically-engineered sushi restaurants employing
robot chefs. *Nowhere* became nothing new. *Ridley* was where all the sane,
normal, un-plugged, under-educated, self-reliant, unreliable, self-starting,
lazy and funny people went.

Sourse was a scary mystery.

None of it lasted long, for Congress, within a year and a half, voted
to annul the sales agreement. Most of the citizenry breathed a sigh of
relief. Some of the new nation states, however, insisted themselves forward, as the courts, international and domestic, were clogged with civil
suits brought by the new owners of record, suits that would last as long
as Mickey Mouse's copyright.

Dr. Fil woke up exhausted and with a headache so large it wouldn't
all fit in his brain, some of it was in his hands, on his feet and in his
breast pocket. Outside it was raining hard.

Ash did the legal work to have Bart Prussian transferred to a security detail on the island of Guam. Doobie Lexicon paid his legal fee by trading one of the barn piles and a 1951 aluminum clad Spartan travel trailer in pristine original condition but no longer street legal. Lawyer Ashwan Clevalure and Doobie Lexicon became fast friends, so much so that later Doobie would name his canyon salvage yard "Ash Garden."

Bob Roy Jeb had finally found the perfect disguise. He had purchased, at state auction, a retired forestry pickup. And at the recycle store he had purchased a forest ranger shirt. It was a tad bit too small, but his observations had shown him that was how they were worn. Now he just had to find one of those Rough-rider calvary hats and everybody would smile and wave whenever he went by.

Enno's land was in geological convergence where the boundary lands of the Columbia Basin blend into the Great Basin of the high desert. The floor of the forest at Snake Flat was waiting for the rains to quit. Waiting to bust forth with small Wild Buckwheat, Wild Aster, Desert Hawk's-beard and Penstemon blooms. These to be followed by Johnny Jump-ups, Indian Paintbrush, and Sand Lillies. This was the season Titus loved best. He would go to the pine forests and sagebrush lands and lay on the ground gazing through the carpets of bloom. He couldn't do that now. As a ghost, when he would try to lay down to see the flowers up close all he could see was the future and that just made him more anxious. He needed to get back there, somehow. He needed to fix things. He couldn't let go of the fact that he might be able to.

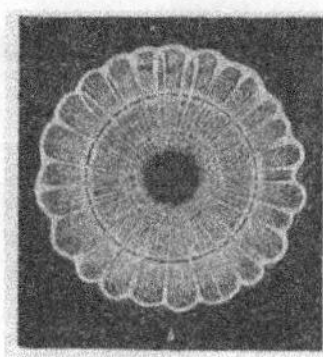

Chapter Twenty

lava rock

Gardens modeled on forest
Positively Hungarian
Sweet sour soar
A French soundtrack
Worried and reaching for
Mediterranean rains
Arabs blowing horn
With Miles in mind
Tiny butt big belly
And Vermillion 'neath the thumb nail

L ook, won't you?"

"In horse days, every cowboy or sheepherder knew more about
his surroundings than the smartest pavement traveler now...
The desert is teeming with life, all engaging in the starkest
drama known, the fierce struggle to live. It has been said that
evolution is merely an unceasing attempt to find forms of life
that can fit into every cranny of this earth."

- E. R. Jackman, 1962

We mustn't just rush in. If agents have surrounded all of the principles, we have an opportunity to watch them watching. We stay back for a day or so. My hunch is that a lot will be revealed to us."

Shirley and Monk were at a gas station in La Pine. She was anxious.

"Ok, but I need to get my eyes on him real soon. Why don't we use one of your contacts and see if there's a way to blend in?" Shirley asked.

At the next pump over, a recycled yellow school bus with windows down held thirty hippies, one with a megaphone.

"We are getting close now, eyes open, watch for him. He will be made known to us." Then came a chant, "E. D. Enoch Drub, E. D. Enoch Drub."

Monk and Shirley looked at each other. She shook her head. He got out of her pickup and went over to talk to the megaphone man.

"I told them we saw E.D. at Weed. They're turning around. You are right about your gut feeling. We need to get there now."

Bob Roy Jeb, in costume and insigniated truck, watched the bus and Monk from across the highway. He followed the bus.

⁂

Filson Wool played Pinochle to win, he took no joy in the game otherwise. He had played just one round with Olivia as his partner and when she insisted on bidding against him, he threw his cards down and angrily stomped off. Game over. Later she explained to him she was holding a thousand aces.

"So? All I was lacking were the aces of diamonds, you could have filled me in!"

"You're an idiot. Forget about card games, the hand you are dealt in this life is your only one. When I came along I allowed you a mulligan, now you're out. I may be too smart for my own good, but we are both idiots and you're still a loser."

✦✦✦✦✦✦✦✦✦

Only those in true service to an unselfish partner beat the odds.

Helga was right, he knew it. Still, it was going to be difficult to have that conversation with Enno. Perhaps the urgencies would absorb it all. This time of year, for Sloop's ranch/farm the urgencies were governed by length of day, weather, and increasing warmth. Throw in the need now to circle the wagons and it would only get more intense. So Sloop would focus, would try to do as much as he could on the farming while keeping an eye over his shoulder at coming storms.

It had, for the time being, stopped raining. Today he had promised himself it was rock picking time. He had plowed and worked up five acres for planting and, as with each opening of pasture to cropping, rocks would work themselves up. He would drive his team and stoneboat around, picking up, or rolling on, stones and hauling them to the field edge where they would be added to the fence line.

He thought, 'Stones and Rocks, know the difference? Insist it matter. Stones be those usually round ones which seem to be of and from themselves. Rocks are obviously torn from or broken off of a larger permanence. Stones pile, rocks might be stacked.'

At the gate he whistled and saw his horses raise their heads. First a walk, then a trot, they came straight away. He haltered each, first the younger mare, then the older, then snapped lead ropes in place and lead them to the double tie stall. Hay was waiting and they started eating immediately. He looked over each one quickly to see if there were any marks, cuts or problems. This he learned the hard way when once he had harnessed a gelding and hitched him, only to find a deep wire cut at a fetlock, impossible to miss when he bent over to hitch the tugs. The mares were clean.

Next he leaned his ear against the belly of the pregnant Percheron and whispered "Hello in there, can't wait to meet you."

As he curried and brushed the team he let his mind ooze out to the new problems; old ones really. The last time around he had brought it on all of them. This time it was something else. Enno was attracting a lot of the wrong sort of attention. He remembered Titus telling him that whenever he hit his thumb, like with a hammer, and it darkened from a blood blister, he knew it would now be a target attracting more blows. That's what was happening here. All that had happened had made of Enno and his neighbors a target for more trouble.

He put the collar on Elsa and went to get her harness. He ran his right hand under the brichen, forward and under the back pad, and reached for the top of the right side hame. With his hand on the left hame, he spread the harness and laid it over the mare's back, pushing the hames into the collar grooves. He pulled the brichen up to the hip, to hang there for now. At her throat he positioned the hames and fastened the bottom hame strap tightly. Then he moved, methodically, to fasten the belly band, the pole strap and (after adjusting the brichen) the quarter straps. This process was duplicated for Queen.

As he was getting the lines ready Helga came into the barn. "Jackson called, said he heard from Lloyd and that all was good. Figures to be back in a week or less. I will have a hot lunch ready at noon."

Sloop bent over and kissed his wife. Then he backed the team out of the stall and ground drove them outside to the flat stone boat, it was a heavy flat steel plate with chain yoke. This was an old style boat with no tongue. Its entire surface drug on the ground. Being so low it was somewhat easy to roll large stones on to it. Sloop had laid a doubletree ahead of the stone boat, it had a heavy hitching hook at center.

He drove the mares to position and bent over to hook the inside tugs. Then he heard something, like a door slamming. He stood up slow and looked around, nothing to see. Then he heard it again, coming from the roof of the blacksmith shed. It was a piece of roofing tin. Two of Helga's guinea hens had jumped down from the tree branches above and slapped the loose tin. Sloop sighed as he finished hitching the team. He spoke to them and they walked off quiet, he walking alongside the boat at first. Then, sensing all was good he stepped on. It was a balancing act to be able to stand and not lean back on the lines. He and his work horses had several bargains, one was that he would always treat their mouths with complete respect. No pressure on the lines unless it was called for.

They walked across the pasture grass to the tilled plot. Bending over to pickup a rock, Sloop thought, 'I need this work, just wish I wasn't so nervous.'

From the house porch Helga watched her husband and thought 'he needs this work, wish he wasn't so nervous.'

The air was waiting for bullets, for excuses, for lunacy, for sulphur, for screams, for crashes, for explosions, for those tiny misleading moments of solace. The air has never been on the side of justice or fairness, or profit. The air only wants what's next.

❦ ❦ ❦ ❦ ❦ ❦

Justice, they will tell you, is designed to protect the weakest among us. That is not so. Justice is designed to allocate and punish while it protects itself. The allocations will go to winning arguments, not to weak people. And the punishments? After the dozen or so inexcusable heinous monsters are given up - swallowing their own blood for the evening news, the punishments go in bulk and wholesale to those with no argument. The punishments will go to the weakest amongst us. And justice will live another day for the next set of lies and mangled fairness. Such thoughts, such observations, must be made away from a need for 'what just happened' explanation. Such thoughts must hang in the air as our closely held self-admonition.

These things we hold as, dammit, self evident: all men are made unequal by their brethren, who shall and do abuse any semblance of life, liberty and the pursuit of comfort in the name of the fortification of arbitrary and capricious wealth.

gggggIrgggggIrgggggggIrggggg

Smudge, as he drove and listened to Jon, was beginning to realize that he was to become a new form of life, and that he was launched now towards the search for a cranny where his survival might be allowed. Here it was, more than just an indication. People are various, as in various forms of life. Certainly the dishwasher/janitor Cochran was no kin to this Smudge armed now to learn how he might join a battle to save the very idea of farming? And a Lexus driving daytrader or the slow-death pharmacist or the cubicle clerk or the chainsmoking judge - how are they not a different form of life from the person who would choose to tend to nature, nurture nature, bow to nature, and explore the sexier aspects of ways and means - enjoined in procedural buffets of tools and practises and appreciations? How are they not different from one who chooses to be a farmer? Throw out the better or worse, talk to me about the difference. Talk to me about relative necessity.

Jon was talking, mostly to himself. But he could tell he had a true audience in Cochran.

"My way of seeing this is that there's a small window for getting Enno's farm truly up and going. And it is more important than it sounds like. This is not a desperate move to get one man a toe-hold on a piece of land and a working sense that he might actually scratch a living from it. This is about a holy example of how it might be that we all can live here for eternity. If we don't find that now, then the earth as our home will die and there goes our heaven."

Smudge observed, "What you need are ants. Lots of ants. I mean a whole bunch of people who can pickup rocks and move them, can build things quickly, can gather food for winter. You know what I mean? But I guess that only works if you can find something they want in return, and that they understand it's a short term job. Because it wouldn't really work if they all stayed."

"Hey, Buddy! That's brilliant. You just gave me an idea."

Anger is such globular misappropriation of emotional juice, on one side slimy and slippery and absolutely useless, on another side sticky and gritty and dirty, and on two more sides, when properly dressed and casually rehearsed quite possibly a tool of near immediate impact. But it is nothing but a skin, and when it sloughs what emerges is a reminder of beauty. You can choose to slough it off, or insist on hanging on to it discovering that anger will crack and itch and flake, and screw up cosmetics, and gather dangerous particles 'neath it. What's so funny about any of that?

After gathering the errant stones and rocks from his tilled field, Sloop drove the stoneboat and team back to the barn. This long pasture he crossed was framed on the east side with scab rocks strewn with ancient dwarved junipers that grew in tortured twisting with mysterious angles confused by long-ago wind broken limbs and torsos. First glance they looked to be dead for a hundred years, next glance you notice some young growth worn like a crippled ol' woman might wear a new hat for church. And there was, during dry times, the hidden dusty green fringe of the lichen - old clothes.

With all of the rain soaking into dry wood, rock, and lichen this once reserved and forgettable useless perimeter of failed and failing desert had sprung dramatically to life. The juniper limbs and trunks, to Sloop's appreciative passing eyes, were now deepest black and simmering saucy reds framed by the brightest, even fluorescent seeming, chartreuse green of the living lichen. Scattered under this wild circus of weed trees were volcanic rock scabs and slabs, sprinkled with their own forms of tidepool-like lichen, and shimmering blue/black/silver and rust colors.

For thirty years Sloop had witnessed this beloved transformation so he knew that if the sun persisted all of this riot of free color would once again subside. This rock and juniper enclave was having a party right now, when the moisture left it would return to reserve. But, a look to the southwest and it was apparent more rain was on the way. The party would go indoors for a while, and return for the next riot when the new sun lit the engagements and lichen would sparkle again. Transformation, change, the morphs of life: these things were naturally on Sloop's mind as he fussed with his decision to apologize to Duden.

No way to know what had changed but Titus knew he was wandering, he couldn't seem to think of where it was he wanted to go next - what pressing obligation he had, all he could find that was familiar was a first exercise - so he sent out a message first as a question, then as a quiet scream. Dog? dog! dog!

Moments before, Resumé had tested and tested the shallow swirl of hot water and finally stepped in to find a rock shelf on the edge of the dish. It was SO relaxing. Steam tickled his nostrils as hot mineral-rich waters swirled all around his seated bottom half. He could have sat there, chewing his gum, forever.

Then came the rude force of surprise. First he heard the question "Dog?" and said to himself 'no you don't,' tightening his sphincter muscle and swallowing his gum. What followed was pure craziness. "dog!dog!dog!" caused him to release a series of gas blasts that parted the warm waters and permeated the hanging steam with the very worst odors imaginable. Resumé jumped from the hot tub and ran behind a big juniper stump. Once there, once away from his own stink, he barked back, short feeble 'leave-me-alone' barks.

He shook his head. Those "dog!dog!dog!" thoughts kept at him, like a ringing in the ears, only this was deep in his brain. He had to find where this was coming from and bite it 'til it stopped. He would need help, he went to find Nimbo's giants.

The Riksdag-Bundesburg-Murdoch committee (RBMC) had its head-quarters in Zurich and purportedly concerned itself with the European Union's negotiating power, but that was a sham. If you were able to dig through all the layers of affiliation you would discover that RBMC was a highly illegal governing board for the collusion interests of the four biggest high-tech companies: that computer/phone company + that giant internet search engine + that social network platform/dating service + that mail order shopping giant. They control a total of over 500 billion in assets with a market share of over a trillion dollars. They are collectively concerned about the welfare of every human being. Not the welfare as in life, liberty and the pursuit of comfort. No, the welfare they are concerned about is the frame-work for taking care of everyone's needs (with a bifurcated priority quad-rant - poor on the left, rich on the right). Welfare as in a global food stamp program, global membership health care, global utilities, global transporta-tion infrastructure, global entertainment, global agribusiness, global this that and everything.

Sounds insidious doesn't it? But truth be known the guys that head it up are flaming idiots so totally married to their swollen egos they make the orange president seem like a small town high school music teacher. Their big and crazy goals for the world will never happen for the same reason that cannibal societies had a hard time growing their populations.

It's not a question of whether or not this consortium will be broken apart but instead how we will cope when it happens. What will happen to the services they have provided. As humans have devolved to their current screen attendant/dependent form will they be able to survive when internet shopping, smart telephones, instant Jeeves service, and the internet peeping Tom service all go poof?

But all of that is way out there, anyway you look at it. What concerns our story is what is right here and now. And that is plenty out there as well.

It would seem that the RBMC had, some time ago, selected candidate areas they might invest in and develop for the next great capital of the world. They hired some expensive consultants to help them devise a way to decide where and how to proceed. Two million dollars later they had their operating plan. First: to decide where, each member of the RBMC board would have to submit a list of answers to these questions. 1. If you were to headquarter there, what would the area/region have to look like? 2. Do you want your headquarters city of the future to be diversified economically? 3. Does it matter what country it is in? 4. Are there classes of people you would deem problematic in your new headquarters city?

The answers were bizarre and often repugnant. Suffice it to say that the end result was the launch, three years ago, of a 20 year plan to build the new existential world head-quarters in Central Oregon encompassing three counties with a projected target population of 10 million. Bend, Redmond, Prineville and Mascara would in essence merge into one metropolitan area. It was identified that they would commence by negotiating an outrageously advantageous 'deal' to build massive data centers while real estate options were secured and specific individuals cultivated and then elected to county commission seats. Through their various lobbying efforts they managed to own the state's senators and congressmen. And currently they work to reconfigure the region's water rights. Pipelines and railroads have been designed to connect this new city with the deep water port of Coos Bay, the potato warehouses of Idaho, the casinos of Nevada, Pike Street Market in Seattle and the taxidermy industry of La Pine. They bribed the Olympic committee to name Mt. Bachelor for the next winter games.

It was then unanimously agreed that in ten years the name of the new incorporated megapolis would be Bendzilia.

Chapter Twenty One

shale

*"What a large volume of adventures may be grasped
within the span of his little life by him who interests his heart
in everything."*
- Laurence Sterne, Tristram Shandy

Ash said, "With this weather, it might be a good time for us to go to the Wilkens farm in Siltcoos."

Enno was a bit lost, he'd been looking for Resumé and had stopped by Nimbo's trailer. "Yessir, I did see him go off with the boys. So glad they are getting along. But truth be told, he didn't go off with them, he led them. I'm sure they'll be back soon." They arranged that

Resumé would hang out at the old sheepherder's camp while Enno and Ash made a day trip.

"We can take my New Yorker, it's possible to make it there in five hours. We should be back comfortably tonight."

A few minutes later Jon and Smudge met them at the forest service turnaround. After introductions Drummer asked if he could show Cochran the farm building site. Looking around at the desert forest landscape Smudge seemed crestfallen.

"So, is this a real farm?"

"It will be one day." offered Duden who seemed to warm to this new man's sincerity.

"It doesn't seem like farmland, not like Nebraska, where I remember."

Jon said, "This farm will be small and have chickens, pigs, bees, a few cows, a team of mules. There's already sheep grazing. Enno has some exciting ideas about combining Peruvian and Chinese farming techniques."

"Mules?" Now Smudge was excited. Enno smiled.

It seemed the perfect time, so Ash offered his surprise. "I have a client who has paid me with a barn in a pile. Enno, if you are interested I would happily gift the barn to you. It was taken apart over by Prineville. Its a western style forty foot by eighty foot with the center open to the rafters for hay. Each side was used for livestock. It's a very basic structure, nothing fancy. Might be a big job to put it up but Doobie says he and his crew of teenaged boys would like to help.

Drummer grinned wide thinking of the idea Smudge had given him. "I got a way we could put it up lickety split."

"My gosh, could I help?" jumped in Smudge Cochran, former dishwasher.

Again Enno felt himself spinning, slowly, choreographed this time to the music in his head, Fred J. Eaglesmith and the Flathead Noodlers singing 'Small Motors,' He was lucky to be amongst friends.

Margaret Curve was aptly named, not because she was particularly curvaceous, but because she always had several angles she worked to conceal. She was a lone wolf attorney who connived to represent every side of any equation she saw benefitting her. Today's day job for her was representing several non-profits working to preserve open spaces in Central Oregon. Her night job was working for RBMC plus three wealthy families anxious to assuage their guilt stemming from the construction of unnecessarily large summer trophy homes on critical farmland they subdivided. She navigated their concerns by securing large donations to the land use non-profits (who in turn signed off on their proposals). And she devised the language for zoning variance proposals that went before the county commissioners, two of which she helped get elected at the behest of RBMC. Margaret Curve, the snake in our story. Late to the tale, but not really - she has always been there, eating babies and gathering tokens and chips.

It was she who gathered information on the ranchers, Ogdensburg, Shoulders, Fork, and Duden, feeding same through her bosses in Zurich and back then to Bob Roy Jeb and Vic Nib. It was she who, through a friendship with Olivia Perchance, had gathered information on Mascara lowlifes who could be of use in clearing the playing field for the 'inevitables.' It was she who cultivated a first name basis with Gunny, Filson and, lowest of the low, Mascara's sheriff posse.

They had offered for Benji to come with them, but he opted to stay behind. He had some things that needed doing. On a trip to town some folks had commented on his old Indian motorcycle and wondered if he was connected with that gang that had been involved in the altercation at the market parking lot. With a little research he had learned which direction they had gone. On a whim he struck out west of town. When he got to Jack Lake Road, they all came roaring out. He pulled over on the roadside. As they went by him Ulrika took a long look and then whispered to Tronk. In a pack they stopped traffic as they made a U-turn and pulled in all around

Benji who sat on his bike as calm as a hidden Mallard hen, setting on eggs in tall cattail grass. Ulrika slid off the rear of Tronk's bike and walked up to Benji tilting her head as she took his measure. She was six foot tall and he four but somehow it didn't show. She smiled a practised Davey Crockett smile and asked,

"Can I have your bike?"

At first Benji said nothing. She continued to admire the bike as he took measure of her. Not her physique, or person, more like he took measure of the space she occupied and saw himself in it.

"Can you spare a few hours?" he asked.

The smile turned genuine, "What?"

"Hop on, ride with me."

Now she saw him, he reached deep inside of her. It was the first time she had ever been able to look into someone's eyes and feel a time-stretched landscape.

She shook her head to clear it, "You're a goddammed dwarf."

And Benji said, "Funny your size doesn't bother me any."

"Shit'" she said stringing it out like a slow exhale.

Tronk stepped up, he had his length of garden hose.

"Back off honey, I've got this." And with that Ulrika sat down behind Benji on his bike.

He whispered back to her, "You'd better reach round me and grab the front of my seat between my legs, if all you grab is my waist you aren't gonna be able to stay on." She chuckled and he started his bike slow and easy. Her gang started up behind him. Within 10 seconds Benji had the Indian maxed on the speedometer at 60. Ulrika grabbed the seat between his legs. He nodded and opened up the bike. It did not seem like the bike's tires were touching pavement. She couldn't remember ever having gone this fast. She ducked her head towards her armpit to get a reprieve from the wind and noticed that none of her gang members were behind her. They had left them back at the chain up area. At the base of the climb by Suttle Lake there was a patrol car with a radar. It pulled out behind them with lights flashing. But it proved nothing because Benji was far far beyond and uncatchable.

"He'll radio ahead you know." And on impulse she kissed his whiskered cheek. She released one hand from the seat and wrapped that arm around Benji's torso, it felt like it belonged there. The other hand kept hold of the seat. He nodded and kept on going. Over the pass he pulled into the Clear Lake entrance and slowly descended to the boat house.

"Let's take a ride."

"You don't call that a ride?"

"A boat ride." he went in the boathouse and rented a row boat.

Ulrika was looking all around at the beauty of the high mountain lake and smiling like a happy child. When they got in the boat, Benji said "you sit in the middle and row, ok. I'm a goddammed dwarf and my feet don't reach the pedals." Getting into the boat Ulrika started to laugh and couldn't quit. She had never laughed so hard in her entire life. She couldn't remember the last time she felt so naturally safe.

She rowed them out to the middle of the lake as he instructed.

Joking she said, "What? You gonna throw me overboard?"

"You might think so when we're done." said Benji with a straight face.

It was a day of break up skies, alternating between flashes of sunlight and clouds that hide. He took the oars from her hands and pulled them into the boat. Then he looked over the side into a lake so clear you could see the tops of three hundred year old dead fir and cedar trees that were easily two hundred feet tall, all under water. Weaving between the trees were rainbow trout.

When she bent over to look down the sun came out in particular force and played games with what she saw. First she saw her reflection, then she saw the tree tops and fish, then her reflection again - only this time it wasn't a reflection. It was her looking up at her. She was down there, fifty feet below, floating and looking up at her while she sat in a wooden boat just across from this brown dwarf.

Then the underwater version of herself turned and Ulrika could see she was holding the hand of a little underwater man. Down there fifty feet was another version of her and she was with this same man. Her whole body started to convulse, and she started to sob. Benji moved to her side and held her.

When they got back to the land, they sat across from each other at a picnic table. She was still crying.

"I remember promises but I don't remember what they were."

"It's enough to know that they are kept now."

She leaned across the table, took his face in her hands and kissed him over and over again, sobbing.

"Yes," he said, "you can have the bike back now. It was always yours. I was just keeping it for you."

"I don't want the damn bike, I want you."

"I've always been yours, always been your goddamned dwarf. Always will be. I was just keeping me for you."

> *I love you without knowing how, or when, or from where. I*
> *love you simply, without problems or pride: I love you in this*
> *way because I do not know any other way of loving but this,*
> *in which there is no I or you, so intimate that your hand upon*
> *my chest is my hand, so intimate that when I fall asleep your*
> *eyes close.*
>
> *- Pablo Neruda*

part three and a half

boca arriba
face up

While we are asleep in this world, we are awake in another one.
- Jorge Luis Borges

Cristina Branco singing Trago um Fado,
the marmalade from oranges of Seville folded into fig jam,
a twelve string guitar vibrating over white Greek beach sands,
Ouzo, goat's milk and thinly sliced roast lamb's quarter,
Sourdough bread ten seconds from the wood fired oven,
Gustaf Sobin alone with his moist sad sentences,
the leather on the belly of her tight vest crackled with broken promise,
Cork trees daring ants trespass,
All of it alone with you.

The duende.... Where is the duende? Through the empty archway a
wind of the spirit enters, blowing insistently over the heads of the
dead, in search of new landscapes and unknown accents: a wind
with the odour of a child's saliva, crushed grass, and medusa's veil,
announcing the endless baptism of freshly created things.

- Federico Garcia Lorca

chapter twenty two

rare animals

No intelligent idea can gain general acceptance unless some stupidity is mixed in with it. - Fernando Pessoa

The old fat man's knees triggered and popped as he tried to walk though the perfume section of Nordstrom's. Every globular fiber of his obese person emitted high pitched squeaks, audible only to nervous dogs, squeals which insisted 'run from this place, escape the laughs, the lookaways, the commiserations, the disgust.' But his legs would over-reach at the knees, momentarily snap in temporary lock and then pop to release. His attempted run was like that of a dry-jointed robot. His eyes sunk further until fear dissolved their sad edges.

Fear breeds idiocy.

The law of the land is a humorless joke wielding terrible sway, it's mechanics a justice system wed to the rules of the game, enervated by the coming tally, up to only the day's needs.

Denmark is still smaller than West Virginia.

There's the Parliamentarian form of government or, as in the U. S. the Pileo'meantwell form of Governance which was soon to morph to a Partiallynormal form of botherance.

Enno traveled with Ashwan in the attorney's twenty-year-old Chrysler New Yorker. It was like sitting, full comfort zone, in an old folk's living room on recliners while the world zipped past you. Ashwan, whether he drove alone or with anyone, needed to talk or listen. Otherwise, after so many long years of remembering his wife and searching after her, he found himself easily slipping into a trancelike daydream that made driving problematic. He began by asking Enno about his mission to setup a farm.

"What made you decide to build from scratch instead of finding an established farm you could acquire." The blank look in the young man's face immediately answered without words. "Silly me, of course, where would any young person of limited means find that kind of money?"

"Yes." answered Enno, and he found himself launching with easy release into a long and compelling narrative. which began with: "Have you ever felt like you belonged someplace, felt it so strong, felt it all the time, like you were some sort of lost family pet who knew you had to get back, no matter what it took, you had to get back?"

"Actually, I do know that feeling, but not as in I have to get back, but as in I have to find. I lost my wife, my friend, my life partner in the woods one day. Never found a trace of her until recently. And, though I feel somewhat at peace, I don't know what to make of that." Clevalure was all poised to go into his own long story when he felt Enno interrupt him...

"I have always felt that if I kept that dream of building a farm in my sights it would happen. I would find it. I didn't know, I couldn't have known, that my friends Sam and Nettie would make it possible for me to have that land. But it happened, and I didn't have any money. It was supposed to happen. Oh, I don't mean that I was supposed to get it. That would be me thinking I had it coming. It's not like that. It's like being in the right place, with me, with my thoughts where they belong, when the possibility asks me to join up. Does that make sense?"

"I think so."

"The business of agriculture often leaves me feeling cold." Enno continued, "I always feel warm and close to finding my place when I am thinking about, or near, or in, a true *farming* world. One where the lives mixed with nature make any dollar talk seem vulgar.

"Have you met Sloop and Helga Ogdensburg? Their farm is over to the east about four miles as the crow flies. You have to drive twenty miles to get to me, but really we are neighbors. I used to work for them. And that made complete sense. Everything about their farm, how they do things and, perhaps most important, why they do things, filled me with that warmth." He paused with head down. "I wish things had happened different. Sloop thinks that I had something to do with the death of a man who was like his brother, a man who was like a father to me."

Ash felt himself measuring. In the presense of this pile of information, coming from one who had impressed them all with his silent resolve, he felt he had stumbled into something quite private. He needed to earn the view, so he shared some of his privacy, but carefully, leaving an important and worrisome piece out.

"Something happened to me just the other day that had a profound effect and I feel a need to share it. I trust it won't seem too bizarre. I had gone to a favorite meditation spot on the creek, feeling more than the usual longing for her. After a good long while I noticed a ways off an old man sitting behind a pine tree. He was sitting with a dying cougar."

Enno looked up.

"I know, strange, huh? But let me finish please. This man was stroking the cougar's head which lay in his lap and seeing me he said, 'Ash, Jenny.'

"I had never seen this man before, didn't know him. How did he know my name. How could he have known my wife's name was Jenny? Then he said "she loves you" as he stroked the dead cougar's head. He was telling me, I am sure of it, that the cougar was Jenny and that she had died, again."

Ash started to weep and had to pull over. Enno put his hand on his shoulder and offered to drive.

"I don't know how I know, maybe because it has always been right there in front of me. The universe is more than we suppose." Enno drove and after a little while Ash fell asleep.

Dreaming, Ash came to understand. Enno was the oldest amongst them. Twenty-something and still the oldest. And he had no past life. This was his first and only life. And as specific as this life was, with the farm dream and all, Enno floated above. Ash understood that when he woke, this was not something he could idly share with anyone. It would not stand the test of the common.

At the restaurant, the entire staff grumbling about the absense of Smudge, three young people in expensive hand-me-down clothes (Territory Ahead, Ferrigamo, Ripoffeos) were in a booth talking. Directly across from them, sitting at the counter was a tall, used-to-be-strong, man in his early sixties, Bo Coin realtor at large.

"The land around here is odd, beautiful but odd," said the red-haired trust fund baby.

Ah, such sweet sweet music to a realtor's ears. Obvious they weren't from here. And even more obvious they were looking at land. He ate his scrambled eggs while posting his ears on command.

"Air is pure mountain, water coming from the glaciers, and all these rich people getting more and more nervous about where their food is coming from. It smells like it's ripe for a new CSA." Said the other guy.

The young woman, angry from the hairline down, shook her head and not a single one of her tattoos moved. "You aren't going to distract me. I'm here to find out about that man, the guy about farming, Enoch Drub."

"Yes, Jacarinda, us too. But like I said, there's got to be a reason someone like that would choose this country to build a farm."

Bo Coin swivelled his counter stool and interrupted, "Please forgive me for interrupting, I overheard you say you are interested in farm land around these parts?" Glancing down he knew those dirty tennis shoes on the girl had originally cost $800 in that shop on Fifth Avenue. "Maybe I can answer a few questions for you. Let me introduce myself, I'm Bo Coin."

"Oh the realtor, yeah, we saw your sign, Coin Brothers." Bo did not catch the wink between the young men. Very quickly they led the conversation into questions about climate change, politics, and organic foods. They were trying to get bits of information while leading on the realtor more than a little. But Bo had been through the school of the ill-dressed, over-educated, grooming-challenged, entitled. They could not hide from him the fact that they were lint in very deep pockets. But neither side could avoid their slice of the culture wars either, and so it was that all agendas were forgotten when they launched into definitions of reality.

"Legitimate farming requires money for suitable land, money for the right equipment and money for lots of chemicals. All that translates to commercial farming being situated near lots of other commercial farming which in turn is located where factories and large populations will support it." These were all things Bo had said in several realtor convention keynotes. Arguments pointing the land preservationists elsewhere.

"Dude, you aren't part of the problem, you ARE the problem," stabbed Jacarinda. "Only thing I need from you is the address for farmer Enoch Drub."

Before he could catch himself from thinking out loud Bo answered, "I don't know of any Enoch Drub farming around here. There is that boy trying to build a farm in the woods, Enno."

"Where is he!" Like a snake, she shot both hands forward and grabbed the collar of Coin's commemorative Pebble Beach ProAm Nike-made golf shirt. Bo recoiled and found his presence of mind, his sort of mind, and got up from the table saying, "Don't know and don't care."

To himself, as he paid his tab, he muttered 'too far gone.'

Later that day he put in a call to Margaret Curve.

Enno and Ash walked through the long equipment shed at the Wilkens' farm, the land from which had all been leased out to neighbors. Sliding door unlocked, they entered and found that the clear plexiglass roof panels let in lots of light. Inside was all of the Wilkens boys farm implements in pristine condition. Right in front of Enno was a Big 4 John Deere hay mower with a six foot cutting bar that had been sandwiched inside two tri-panel corrugated cardboard sheets. Peeling back the corner Duden could see that the sickle sections were moist with grease and looked to be brand new. Down the row he saw a seed drill, corn and bean planter, cultivator, two plows, a disc harrow, a spike tooth harrow, a potato digger, an old John Deere model A tractor, a side delivery rake, a 14T JD baler, a McDeering loose hay loader, a hay wagon, a silage chopper/blower, a roller packer, a belt drive buzz saw and a chain pasture harrow. Tucked away along the walls were eveners, wheels, spare tongues, and all manner of parts and pieces.

Ash said slowly and carefully, "If you wish, this can be all yours. They left me to decide who I might want it to go to. And I say right now, this ought to be part of your inheritance."

"Inheritance? How could I..." Most of the time Enno didn't feel himself breath. This time, though, he did. Slow, deep breaths. After a few minutes he nodded yes with deliberation. "But, I don't have a way to move all of this. And it looks like I should keep it under cover."

"I was left with the means to move this equipment wherever it needed to go. And it can stay here until you're ready. That barn we were talking about? It's big enough that you could park all of this on the one long side. Now, I have something else to show you."

Lloyd is in Cy Twombly's former studio. The light is perfect and the paint splattered grey board floor works for him. Ceiling is a little high and it's hard to heat. On the drafting table he has three sheets of good Rives BFK paper, each has a similar design drawn in walnut ink and charcoal. Each drawing is bordered by lots of cursive notes. Ever since Cash had explained that the contest might result in a statue for Central Park to be titled Civility, Lloyd had a clear vision. It was the boy Duden who had taught him about civility. A boy forty years his junior. At first his sense of Duden was that he was stand-offish. Kept to his own thoughts and deeds. But with time, and the swirl of recriminations that came out of the eruption and Titus' death, Shoulders had come to understand that Duden had an old-fashioned western sense to him, a self-preservation notion that says "I won't pry into your business so please leave me to mine. And the business I choose is farmwork. See me work. See what I care about. See how it feels. See how it hurts no one, helps everyone, enriches and warrants." Enno was the epitome of civil.

From there, thinking back on the boy/man, Lloyd remembered a particular vision of him at Sloop's, cutting some weedy oats with a long old scythe. That came to be translated to a sculptural form of the Icabod Crane physique of Duden walking and swinging that long knife. That form was molded from found limbs of juniper and oak worked together with hand forged steel elements, plaster, and wire. The scale was important, It had to feel enormous while actually being life size. The torso was to be like a folk ballet dancer save for the shoulders that employ a wound posture to swaying rhythm and the scythe blade to mirror the working curve of his body. His head was to be forward on a borrowed neck, the front foot solid, the back just leaving. This figure of a man was to anchor to a steel plate eight foot by eight foot, covered in a bed of twenty penny nails. Off the back side of the scythe blade, falling in a retreating cascade like the long hair of a joyously loping first-day-of-summer woman was a bouquet of brass wire each parallel strand topped by strings of tinted glass beads. It was to be the harvest of Civility. The shape, form, attitude was all of it in absolute allegiance to Giacometti's relay.

Shoulders finished in three days. Apercu was thrilled. A press conference was scheduled with video to run on the NYC arts and culture venues. But he couldn't find the artist. He found instead a note.

"This is the proposal for the sculpture. I am going home now to build it, whether you like it or want it or not. I don't want your foundry to build it. I want my hands and heart to leave stains directly on it. When I am done I will rent a suitable storage unit near the Redmond airport and put it in there. If you want it, send those funds to this South Sudanese hospital. When they confirm receipt, I will send you the storage address and the key. Perhaps most important, you and I are done now. My remaining years will wash from me as I choose. There will be no more sculpture.

Lloyd was at Washington park kicking trash, and walking like every destination he had ever had was behind him. It was time to go home.

Cash read this and understood that he could not trust Lloyd's retirement from the arts. He needed public evidence of the good cowboy's death.

Here's the place where you, the reader, may be so heavily invested in these stories that you would do well to allow yourself to add your own.

The old man forgetting, forgetting, retreating. He was alone and he couldn't get his sweatshirt off. He'd get it part way and cover his face and tired arm muscles would quit. He had to sit that ridiculous way for long minutes until the strength returned enough for him to pull the shirt back down off his head. Finally he went to the bathroom, got a pair of barber scissors and cut the shirt off himself. His tears rusted the hinge of the scissors.

Pausing he remembered when people silently insisted he hold still so they could look at him, look a long time. They inhaled his beauty. Never presumed, never touched, always carried away light with them and became lighter. But it only happened to him between the ages of 30 and 50. He didn't know what it was. Didn't want to know, just walked around as though in a borrowed body. Things came to him, things to say. But to claim them and offer them up was to kill and he wasn't ready.

Now as an old man who had to cut his clothes off, no one wanted to look at him. He was wasted. He sought a last good use of himself.

He saw the proud, magnificent head of the superb black actor Steve Harris facing him and he knew there was work still to be done. The hearing aids, not always comfortable, did allow him to hear the Red-winged blackbirds warn the Goldfinches off the sunflower seed feeder, that one, on the other side of the kitchen window. What fun! To hear the birds click and pop at each other. To effortlessly eavesdrop on the audio texture of a world he and she so cherished. He shook it off and Harris went poof back to the old man's imagination bin. NO. Today he had promised himself he would tighten his focus and find that one last good thing he would do to repair a lifetime of indifference.

On a corner of the Wilkens' farm stood a faded white clapboard church, looking at first glance like an old one room school house. It was surrounded by waist deep Canary grass. There was a foot path to it. You had to park some distance and walk in. Running his hand over the tops of the grass, Ash cursed beneath his breath. "They were supposed to keep this cut. What a hazard."

Enno drank it in. It felt right. Once inside the building they were surrounded by twelve foot high book shelves tickled by shafts of light from the two pair of opposing skylights. Ash was reassured, it did seem like the ladies had kept it dusted and orderly. Two library tables, two over-stuffed parlor chairs, an old, flowered sofa, and a rolling warehouse ladder were scattered about. It was indeed a cabinet of reading, as the Wilkens boys had called it. Glancing around it was obvious that most every volume had something to do with farming, homesteading, home crafts, handmade projects, etc. Enno went to the first bank of shelves and reached forward for a book on blacksmithing, stopping just short and turning to Ash.

"Yes sir, that is why we are here. You take your time. If we are to be back today we have about an hour and a half to spend here. If you have an interest in this, I will make arrangements to have it moved, building and all, just like it is, to wherever you choose."

Ulrika dropped Benji and his backpack off at the Alice Greeley Trailhead parking lot. He would walk cross-sountry to Snake Flats. She would take the bike and finish her business. If they were to be together again, it would be...

"Well, art is art, isn't it? Still, on the other hand, water is water! And east is east and west is west and if you take cranberries and stew them like applesauce they taste much more like prunes than rhubarb does. Now, uh... Now you tell me what you know."
- Groucho Marx

chapter twenty three

early blooms

We could use up two Eternities in learning all that is to be learned about our own world and the thousands of nations that have arisen and flourished and vanished from it. Mathematics alone would occupy me eight million years.
- Notebook #22, Spring 1883 - Sept. 1884 Mark Twain

"I have a strong propensity in me to begin this chapter very nonsensically, and I will not balk my fancy.--Accordingly I set off thus:"
- Laurence Sterne, The Life and Opinions of Tristram Shandy

The French baker, the Greek fisherman, the noodlemaker Sam Wo, and a Jersey putz went into a brew pub in downtown Bendzilia to meet with Hardlee Nuff, the inventor of RiffRaff, the new atonal scatpap-chat-boogie taking white college ghettos by storm. They were there to discuss a concept for what Hardlee called an 'whoneedsadayclub', a place where, 24-6, you could get a stiff drink, a fish sandwich made with a fresh croissant and a bowl of pizza-flavored ramen while watching daytrippers jump and flop-around to RiffRaff. One random day a week the club would be closed. Hardlee thought this would teach the creeps not to take the place for granted. Hardlee needed to raise twenty-five grand, the going rate to bribe the county zoners. The baker threw in a thousand euros, the Greek signed over the pink slip on his trawler, Sam Wo wrote a check for 20 grand and the Jersey putz passed. All but the putz knew a good thing when they smelled it.

> *"It's a shame that the only thing a man can do for eight hours a day is work. He can't eat for eight hours; he can't drink for eight hours; he can't make love for eight hours. The only thing a man can do for eight hours is work. "*
> *- William Faulkner*

Faulkner died too soon to witness this age of committed, sixteen-hours-a-day, indolence and welfare obesity.

⁂

At the edge of Snake Flats a gentle rise of sandy bunchgrass ground, speared by 3 groves of eleven each Ponderosas, nestles a lovely box canyon the back of which features a red palisades cliff rising to rimrock. The base of the cliff is shaded by bitterbrush and willow. Within that, at the center of the wall, is a crack entry to a large cave, an ancient shelter to indigenous peoples, the back of which was recently torn open by earthquake to allow access to a massive cavern, wandering with three long arms, joining to form a central hollow within which a Brown Dwarf star burned, in hover, like a endless sparkler display showering out those particles our Dr. Fil has named Filenium. The Brown Dwarf star hung and spun over a three mile wide lake, like an underground spastic sun, providing heat and flashing light. The center of the lake bubbled with deep-core heated mineral rich waters.

On a precipice overhanging the lake, floated Titus Ibid. With him came a wind.

"It's like miracles happen to him, not that he makes miracles happen."

"But, think about it, isn't that the circle of life he represents? He is a miracle, miracles happen for him, to him, because he exists, and he directs miracles."

"Well that last part I can't say I've actually witnessed. Look, he's right here, or over there, close at hand yet it feels like he's very far away, maybe even not of this time. Maybe not of this world."

"That's the hazard. Don't do that. Don't make it more than it is. The transformative example does not need to be miraculous."

"Listen to you. If an example is truly transformative it is most definitely miraculous. And this is not a semantic argument. Transformation that comes because of an example, in this case the man who wants to farm, is change caused by what? Good or bad, it's change caused by a perception of unusual intangibles."

"But none of that is cause for people to rise up and do things."

"It most definitely is!"

Promto Manks, in his electronicized van in the Walmart parking lot, has pulled off another coup. He's followed the digital leads and found their central heating. Four points: Zurich with its RBMC operations, New York City with Apercu LLC, the formaldehyde soaked denizen of the office of Homeland Security in Jackson Hole, Wyoming and finally the office of Margaret Curve of Bend. He encrypts this info, draws a couple of possible conclusions, codes those, and sends it through his new circuitous routing to reach Monk and Shirley as they enter the outer edge of the city on the Deschutes.

At Poor Viet, over two bowls of Bon Bol Wat with plucked duck, they hatch a plan.

After three days of illness and goofy behaviours, and with the help of Hector's wife, Ilbana, Victoria smiled her acceptance and then went into a deep fear. She had to be with her husband, and right now. Hector and his family of ten would care for the Central American rainforest homestead.

But Ilbana insisted she must go with Señora. And so the two women started their trek to find Señor Sig.

Dr. Fil's head was spinning. He forced himself to separate his thoughts. On this side were the discoveries in their simplicity and complexity - oh my gosh Filenium! And on that side were his ideas of what they or that might be made to do - or made to mean. And there was a third arena; the young man and the coterie of engaged companions he had drawn around him. There was an undeniable energy and goalset to all of that. It was strangely contagious. And it was dangerous for him and the cavern, source for the particles. Too much attention could be a problem. So he set himself to wonder if there was a way he could draw Duden and his friends away somewhere else, somewhere better suited for the boy's dream of a farm. That shouldn't be too hard, he chuckled. Afterall, he had Jimmy and, perhaps, the little old attorney.

Jon Drummer had left Smudge with Uncle Nimbo. Smudge was thrilled to be standing around with a long bent staff and watching grazing sheep. He felt like he was a picture from one of those old illustrated Bibles of his youth. Nimbo asked Cochran to help, he needed to try to find his dogs. Whenever they disappeared, and it seldom happened, it worried him. And they had been gone the whole morning.

"Do you know how to shoot a rifle?" He asked Smudge, who shook his head no. "That's ok, if anything threatens the sheep jump up and down, yell and wave that shepherd's crook at them."

"Yes sir." Smiled Smudge Cochran, born again farmer.

Jon went in to get a burner phone. He had some people he needed to line-up for a barnraising. His first call would be to his Amish-raised friend, Joe Ablehammer.

Vic Nib had retraced all of his steps and found his way back to the Italian restaurant in Manhattan where the man who had hired him was said to hold court. Leo Vitch had been close behind him at every step so far, but now he was forced to divert and pay homage to Anatoly Grim. Next would be his hit on Nib. Grim had just been informed by Cash Apercu about the impending permanent retirement of Shoulders the sculptor. Grim was not very happy and informed Vitch that there was a job he had to do right away. As far as Anatoly Grim was concerned, Lloyd Jerald Shoulders was not going anywhere until Grim chose it. Cash was no longer of use to Grim.

Vic Nib had been keenly aware of Leo Vitch, knew he was being followed and targeted. And he preferred it that way. So now, when he couldn't smell the Russian he worried. Monk had taught him as much. Keep your adversaries where you can smell them. Whenever Vic Nib thought of Monk he reached for the nub that was his lobe.

At Mama's Trattoria, Nib had learned that boss Sergio was a relay. The person who had actually ordered the hit on Duden and company was a female attorney, and Sergio explained that she took instructions from Zurich's RBMC. He had gotten the information he needed, but, for practise, Nib decided to shoot Sergio anyway.

"Computer dudes? Who would have thought it?" Nib purchased a one way ticket, NYC to Zurich, with a return ticket to Bend.

Bob Roy Jeb had followed the busload of seekers to Weed, California and learned that Enoch Drub aka Enno Duden (or so he thought) was holding a tent revival in Redding.

Plumley and Recto, realizing that the heat generated by the *Engine of Quanchee* was increasing the market value of their internet holdings, decided to up the ante. They found a psycho blogger peer and coached him to become prophet Enoch Drub. They then hired a promotions firm to put together a rock concert and tent revival on the outskirts of Redding, California, in a wildfire charred field. Social networking was working and every sign was that thousands of seekers would be descending on the site. Plumley and Recto decided they would book themselves to play the concert, in drag.

A California operative had passed along the outline of this to Margaret

Curve's office. That information had come through her computer at the exact moment Sig Maltesta was sitting at her desk. Shirley waited outside as burglar Monk looked for bits and pieces of information. Monk knew a convenience when he saw one. He typed in a reply on Curve's computer, "Do nothing, Wait for word from here."

On the way back from the coast Duden observed, "The equipment and the library are fantastic. But its pressure. And I'm not sure we are ready for pressure yet."

Ash heard Enno say '...not sure if WE are ready....' and his old soul felt accepted.

"These things are there, know that. And you tell me when and what you want done with them. Ok?"

"Yes, that's good. Ok." He paused a moment and then said, "Can we talk about Dr. Fil? You know that my land is the land I choose? Right?"

"I know now."

"When we get back I need to show you and Benji something right away. And then I want to understand how realistic it might be to get a field cleared of stone, that barn up, and some waterways set."

Hearing these things Ash was thrilled to his core.

Shirley was shaking. She was shaking with it. Soon she would see him. It had been too long. Would it still be there? Soon she would know. Monk reached across and set his hand on her shoulder as she drove. They smiled at each other. She nervous, he knowing.

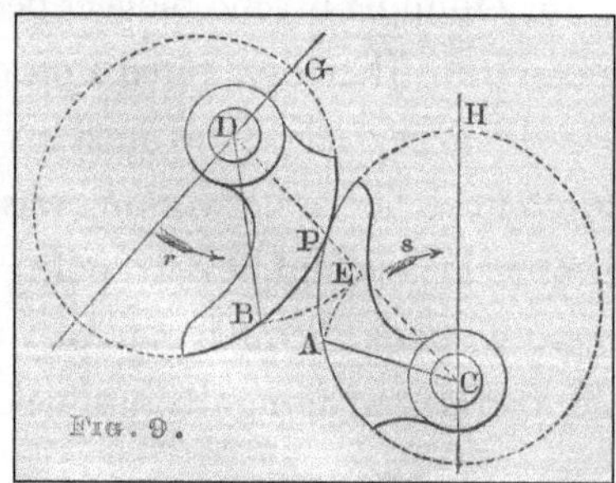

chapter twenty four

soups

more good farming more good farming!

I wasn't meant for reality, but life came and found me.
- Fernando Pessoa

Gendun Drub was born to a family of nomadic farmers in 1391 near Sakya in Tsang. His father was named Gonpo Dorje and his mother Jomo Namkyi. His birth name was Pema Dorje. His family belonged to the Drom clan.

According to legend, on the night he was born, the family's camp was attacked by bandits, and his mother, fearing for the life of her newborn child, wrapped him in blankets and hid him among the rocks before she fled for her life. The next morning, upon her return, she found her son resting peacefully among the stones, with a large black raven standing guard before him, protecting him from the flocks of crows and wild vultures that had gathered to attack him. The raven is said to have been an emanation of Mahākāla, who would become Gendun Drub's personal deity.

 - read by a bald cat at a new-age meditation seminar in 2018

The dogs finally returned, sheepish. (Oh, you must pardon that - livestock predator control dogs described as sheepish? - you must pardon that.) Mikey, deGaule and Sarkozi seemed disoriented and more than a little tired. "Off chasing wolves were you?" asked Uncle Nimbo. "Ok, that's good but you worried me. I almost shot off my flare gun." Nimbo had promised his dogs he would only ever shoot off the flare gun IF he or his sheep were in deadly harm's way. 'No crying wolf.'

As they walked back to sheep camp, Mikey kept looking over his shoulder while making whimpering noises.

Resumé was walking slow, thinking. When the other dogs left he figured he had better find Enno. It had been dark and windy in there. Hard to get a scent reading. But he knew he had to go back. Something was calling him.

When they came into Bend, Shirley and Monk avoided P. M. They sent him a new request and then went straight to their burglary assignment at Curve's office after which they drove in silence to Mascara. Sig thought about Vicki back at the homestead. And about the circles he felt closing. They had succeeded in getting Bob Roy diverted for now. Nib and Vitch had taken Fred's bait and were back east stirring bad pots. The D. C. news sent Colonel South and Grey Water to Mexico, forgetting about their assassination assignation. Maybe just maybe, Curve had been sent on a wild goose chase. But her involvement was most worrisome to Sig. Stopping her was key. Even so, it felt like they had earned a little time and space.

Dr. Fil had made a quick purchase from a colleague of an idyllic farm in the Santa Cruz redwoods, a place called New Hope Farm. He put the farm in Duden's name. The plan was to trade. How could the young man refuse?

When Duden and Ash returned to the farm site, Benji and Resumé joined them to go see the cave with the artifacts. Resumé was plain nuts when Enno got back, and over the top when he figured they were following him right where he wanted to go. He'd go fifteen yards at a run, stop bark, leap in the air, and run back to them only to repeat the ritual.

"He sure missed you," said Ash.

"I think there's something else," responded Enno.

This time they had flashlights when they turned sideways to slip into the crack in the rock wall. Resumé ran in and disappeared, Enno calling after him. Then they reached the outward curving rock and saw the ochre, burnt-red, and black hieroglyphs. There were geometric symbols, animal silhouettes, and people, dozens and dozens of images painted in layers over the top of each other. And, what Enno had not noticed before, on the other facing wall were shelf notches cut in the wall with human skeletons. The bottom shelves were empty or had bones scattered. But in the top shelves the skeletons lay in repose and seemed to be complete. Much of this cave was dusty and webby as though it had been without disturbance for eons. But in some places there was rubble that seemed new.

"I'm guessing this was a tomb or burial crypt that had been sealed off. There's some indication that it may have been recently opened, maybe from that earthquake and eruption?" Ashwan was tracing the petroglyphs as he spoke.

Benji was silent and stood with eyes closed. They heard Resumé bark from deeper in the cave, and then from the same direction there came a breeze that developed into a wind which carried a strong methane odor. Resumé came running to them with a squeal and then, shaking his head, he continued right on past and out of the cave, his stub of a tail pointed down and held hard.

Ash said, "It's time to leave, that methane smell is ominous."

Jon Drummer had made interesting offers to Joe Ablehammer challenging him to a goal: could he marshall a crew to put a barn back up in three days?

"I'll answer that when you answer this: Can you feed a crew of thirty working men, feed them well? Can they camp on the site? Can you keep the county inspectors at bay? Can you keep the immigration police away? Will you be able to get a generator there to charge our portable tool batteries. When do you want us to start? Can I ride with Lady Gaga in her tour bus? If the answer to all those questions is yes, you're on."

Drummer also had contacted a Chicano stone mason, José Jimenez, and requested he bring a crew to lay a dry stone barn foundation in three days.

"Por usted Señor Drum, dos dias, IF it is as you say. My family will be enough to do this, we are fourteen workers and 7 children. But remember if we do this thing you will owe us."

"I understand mi amigo, if you do this for me I will arrange with the Mexican Government to grant you visas to return to your home country. I'll get back to you with a timeline, but it will have to be real soon." The most popular rock and roll drummer of his time was going to be calling in all of his chips on this one.

Jon gave Joe and José the word, Nicodemus.

That done Jon took a deep breath realizing the more difficult work was ahead. He had to tell Enno and Benji that at least 51 people would be descending on Snake Flats for a short week of work. Then he had to get Ash to deliver the barn materials. He also had to alert Uncle Nimbo that no building inspectors or I.C.E. could pass onto the property. And lastly, how the heck was he gonna find a way to keep the crew there and working? Where was the food going to come from? Then he remembered his dishwashing/farm dreamer friend, Smudge Cochran.

OOOYYYYYYYYYOOO

<A stage set for the sermon on the rocks? Just asking, I know this story is moving along without the rudder of a good narrator. Remember me. Whether you call me Dude, or Rude or Crude or Levi Tookus, If I hadn't been there in the beginning... Wait! Don't!....>

XxXxXxXxX

While at the college in Santa Cruz, Dr. Fil came across a new report from friends at the Deep Carbon Observatory. Guiseppe Etiope and the group had recently discovered what they think causes The *Flames of Chimaera*, referencing a perpetual fire pouring from the top of a Turkish mountain. It was abiotic methane, the result of chemical reactions involving water and rocks miles below the earth's surface.

Dr. Fil nodded his head enthusiastically as he read. "Yes, yes," he muttered, "the planet is a living thing, a living thing of its own design."

Now it seems that the exact same phenomenon is occurring in hundreds of spots around the globe, some at the bottom of the seas.

The recipe is not simple, take the hydrogen from the water and mix what remains with inorganic carbon taken from minerals or other gases. Add a metal-rich mineral to trigger the process and you get abiotic methane. The hydrogen in the Turkish case comes from serpentinization, which is the result when that water percolates through mantle rocks. There are variants in the recipe, different minerals, carbons.

Or the hydrogen might come from friction or radiolysis instead of serpentinization. The temperatures involved range from 250 to 900 degrees Fahrenheit. So The *Flames of Chimaera* happen when carbon dioxide rich limestone, or calcium carbonite and hydrogen rich serpentinized rocks are doused by rainwater.

"No, no." muttered Dr. Fil, who asked if he could use his friend's lab for the afternoon.

Looking for exception he came up with the narratives around The *Lost City hydrothermal vent field*; most of that methane, formed under the extreme heat, was the result of once-buried mantle rocks being exposed and serpentinized by circulating seawater.

Though perhaps not, because in each case we see chemosynthesis — the deep earth version of photosynthesis. Its energy comes from a chemical reaction, not the sun. Inorganic molecules are transformed into organic products life uses. It's the basis for all ecosystems on the planet. But for Dr. Fil it was also a theoretical conflict for his discoveries of containered microscopic particle accelerations.

If and when chemosynthesis forming abiotic methane also produces amino acids, the resulting corrections would be damning. The key for Fil was to control rudimentiation. (Yes, not rudimentation.)

<And but, and though, and really now, the key was to curb socialization. The key was to control human society.>

Dr. Fil sat, head in his hands for an hour, thinking, 'What truly happens when radiolysis and serpentinization collide, in an enclosed chamber, with dedicated peripheral friction?'

<What happens when me-first, I-got-this, no-problem, humans mess with that which they may never grasp let alone own?>

Helga had surprised her husband. "John, I want to go with you. I need to see him too." John B. 'Sloop' Ogdensburg loved this woman more each and every ten minutes of each and every new day. And his new discovery was the amplifications that came of loving her through her disappointments with him, in him. Each time he caught himself smiling at and about her and their life together, another set of old man aches lifted and fluttered off.

Driving to Enno's land Helga observed "Ever notice, John, how these days we don't have many bugs on our windshield? I'm afraid, it isn't right. I'm afraid the chemicals are everywhere now. Yesterday when I was out feeding the chickens I picked up that old coop ramp, that piece of rotting plywood? Under it were bugs, lots of bugs, and a couple of earth worms. I was thrilled to see them. Just thrilled. That's how things have changed. And then, in the blink of an eye, the hens saw them too and gobbled them up. My emotions just went on hold. And I realized, that's how it's supposed to be. Not the gobbling up, but the balance and the balances."

"Yes, I have taken to greeting any spiders or ants or bumble bees I see as though they are prodigal sons, as though they chose to go away for their own profit and now have returned. If they chose to go away it was to avoid being poisoned."

With the words *prodigal son*, they both paused, looked ahead, and found themselves reaching together to hold hands. They knew that the prodigal, the wasteful one had been he, John B. Sloop Ogdensburg.

Jon and Benji arrive at the same time to rejoin Ashwan and Duden. Ash finds the right moment to tell Enno about the 1951 aluminum clad Spartan travel trailer he wants to gift to the farm. Enno smiles and hesitates to say anything, thinking about how much he is looking forward to sharing his stone and pine cabin with the woman he loves. How it is he has no interest in the trailer. But that's one of those thoughts he shunts three layers down, just there, nothing to act on. Lots of other things to act on.

For example, Benji. Enno had come to feel his short wise friend to be an extension of himself and the farm. Benji being gone all day with no word, that worried Enno. He realized there had been no conversation about any future for Benji in Enno's plans. Was he to do these sorts of things? To have conversations? Make invitations, utter 'no thank you's? Was he to find right language? Too much to think about. So instead of uttering he went with the moment and muttered, "Thank you, if he wishes, the trailer will be for Benji to use as a temporary quarters on the farm. I'm hoping he will want to build a permanent home here. "

Benji kept his large, infrequent smile deep down inside. He hadn't needed the confirmation but just the same...

Enno found himself looking at the cabin, standing now with roof, porches and two of the windows in. Around the structure hanging from tree limbs were a pair of fine mesh, rabbit-wire cones Benji had made, each filled with black oil sunflower seeds. He had hung them two weeks ago. Above one, a squirrel dangled trying to reach the cone. Two Red-winged black birds were dive-bombing him until he lost his grip and fell to the ground, darting off to a wide dark crack in the old juniper. In a crazy wide halo, Enno noticed the many different birds who now saw these cones as their cafe site. There was a single Yellow-headed black bird, dozens of Goldfinches, towhees, waxwings, nuthatches, grosbeaks, and starlings. On the ground scurrying for the fallen seed were quail and dove. It was hard to imagine feeling more peace and assurance than in this moment.

 Then. He hadn't heard Shirley approach. The smell he was allowing himself to engulf was familiar but he had mistakenly given it to the birds and the moment.

She came to him from behind, hands to his shoulders, down his upper arms. Staying behind him, her hands found his - she cupped one, spooned inside hers, and allowed that he cup her other, spooned again - and the four arms so joined wrapped his belly, her front his back, and the union was made forever.

And no one could watch, for it blinded them all, even the dwarf whose gladness was a terrible force.

In this way God kept his promise in full apology.

All who understood were sworn to their core to protect. Even Sloop, Helga and Monk who watched from just far enough away to have to peer through the glow.

Now facing one another and embracing, Shirley and Enno turned together and walked into the cabin shell.

No one plays high notes like The Lip
I never heard a trumpet player play a note so high
And I had to coax a lot before The Lip would tell me why
Then he took a little jar that's labeled 'High Note Grease"
And he rubs a little every night on his mouthpiece
Yip Yip Yip
No one plays high notes like the Lip
Listen here gal, are you kiddin' about all that 'high-note grease'?
No, man, I swear, he had ten in his valise
Wha', you mean he goes to the drugstore and gets them from the medicine shelf?
No, some cat's told me he makes it himself
Yip Yip Yip
Tell us the secret of The Lip
Well... you take a bucket full of steam
And a dozen rooster eggs
And you mix'em up gently with a bushel full of goldfish legs
And ya hang'em on a sky hook in the midnight sun
Mmm and then you fry them til they're done
Yip Yip Yip
That's the secret of the Lip
                    - Louis Prima

He was no longer a boy, no longer a young man. Man, bull, ram, stag, stallion, boar, none of these genderized satchels of attribute fit him now or ever would, neither was he indeterminate. His was not a human specificity, his was a natural attendance. He was not put on this earth to breed and bust things up. He was put in this world to demonstrate communion, to demonstrate love, to garden in the garden for the garden and with the garden. His was a presence only in his presence, which is to say he would forever be recipient of his chosen world. He needn't have a name, he needn't own, he needn't possess in the 'slow this down, stay away, don't touch it it's mine, if you think it's grand it's because of me' way of most limited and limiting human males. His was the freshbaked poise of the appreciative dog in love with an entire good family, an entire world, and entire patch of spring grass to roll in. But his oeuvre was all of biological life. This new male of this new species was an embodiment of the zones of mastery, outside himself because he was spread wide to touch all things that would touch him. He was not the only one of his kind. He was one of the first, which mattered not, save to say, that these magnificent and terrifying examples of living in life that spring from his, and their, touch are more than options. They are the path towards each next sunrise, each next harvest, each next planting, each next calving, each next perfect rain, each next partnering touch.

She was no longer a girl, now forever a young woman. She left her species behind without regret and joined him to complete him as separate, as partner, as navigator, gunner, nurse, as polisher, as happy hour, as happy day, and grateful life. Hers was the slow-cooked insouciance of the dedicated, indifferent, yet completely possessive feline. She needn't have a name, she owned him, and she was blissfully grateful for their union in absolute devotion to each other and nature. She was the only one of her kind, the woman who all along knew the sadness of missing him, even as he was there. She knew he had come from eons before her, and she knew he would extend beyond her through to the beginning of each new time. What saved her from this sadness was knowing that his was a curvilinear trajectory and that they were together on each and every overlap. She was the reward for the first who mattered and in this way she knew she was the first who mattered. And together they were the first who mattered, and they spoke in one voice but with clarion harmonies that fluttered leaves before the breezes, and alerted fledglings that mama was on the way with food, and caused ants to pause in grateful salute, gave arabesque to otherwise slow clouds, gave to life the tasty oils of a slow stew well set.

Lippenbär.

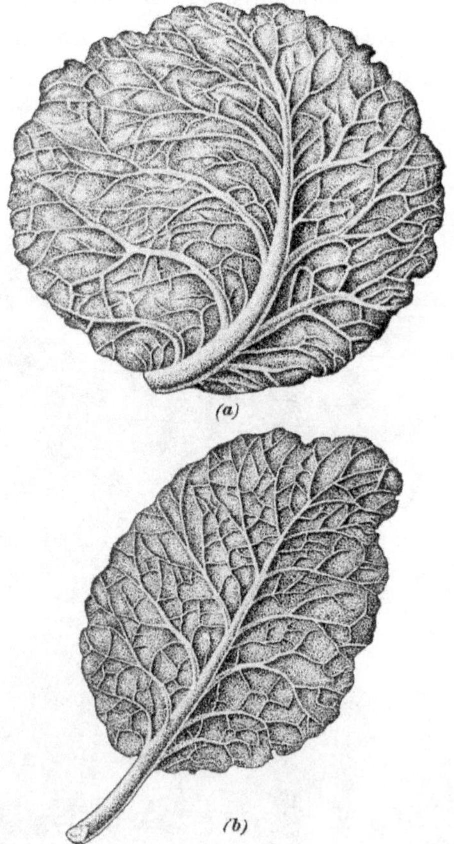
(a)

(b)

Part the fourth

epic epilogue?

the clock has stopped
bong bong
time now at attention
schwing schwang
life now demands
seed stuck in the teeth

"Clocks slay time... time is dead as long as it is being clicked off by little wheels; only when the clock stops does time come to life."

- William Faulkner, The Sound and the Fury

chapter twenty five

barn siding

"Digressions, incontestably, are the sunshine, the life, the soul of reading! Take them out and one cold eternal winter would reign in every page. Restore them to the writer - he steps forth like a bridegroom, bids them all-hail, brings in variety and forbids the appetite to fail."
- Laurence Sterne

Behaviors should be vaporous, actions etch enough. You are what the meat you ate ate. Feed the meat you eat poison and test your life expectancy while you test the circle of life. Learned imbeciles oblivious to the generous truth of the good heart, lead us not to convenience and deliver us from barking cleanliness for wine and brine are the power and the glory.

The Myrmecophilus Group, steering committee for RBMC, were in Bend meeting with their operative, Margaret Curve and Company - Open Space Attorneys, to further intricate the design for the new economic and systemic capital of the world, Bendzilia. Over three of the last four years, heavy late winter snows had returned discussion to the manipulations required to assure that the Central Oregon Cascades would become the near future home of the winter Olympics. (They secretly planned for it to be a permanent home for those exalted games.) Mt. Bachelor already featured a respectable ski facility, Hoodoo less so, but the landscape was there to expand and quickly. Margaret was directed to move ahead, exercising the option to purchase, from the Del Webb company of Nevada, the old railway easement that ran, circuitously, from Mascara to Mt Batchelor. An Italian firm was designing a mass transit version of a high speed aerial tramway which would transport people back and forth in great good numbers.

The Myrmecophilus president, Jay Bossweiner, was making a case to the meeting of how, obviously, once word got out of the master plan, real estate values would skyrocket. And he pointed out that their long term work preserving the wildlife habitats and farm lands intersecting the route had put them in perfect position to now 'trade' those conservation easements in creative ways. Twelve years of purchasing deeds and options for thousands of acres of ranch and forest lands were soon to pay enormous dividends. Margaret sat alternating from beam to frown, she was still worrying about her failed efforts to thwart the messianic triangle while thrilling over what was to come.

Bossweiner continued, "We were originally opposed to it, but now we see the economic value in maintaining a ten floor limit on all construction. Our architects assure us that we can hold to the goal of a massive grid of interlocking 'villages.' Co-ordination continues to be the key, that and continued secrecy. Imagine friends, Bendzilia will comfortably contain ten million in population, with the savoir faire of San Francisco and the panache of Seattle, it will exceed the financial oiliness of London and Dubai combined while being a dryland equivalent of Venice. Every banking and software company of any worth will need to be here, actually and actually."

Next morning, hope being there will always be one, Jon Drummer tells Enno, Shirley and friends about his pulling together two crews, one for the laying of a barn's foundation stones, and another to raise the barn Lexicon and Ash had gifted the farm.

Ash asked, "How did you arrange this? Whose to pay and what?"

"I'm making it happen. No money changes hands. I've worked out trades."

"We need to know what that means young man, there is danger here." offered Helga.

"Fair enough. I am owed favors by significant individuals. I will be arranging that José and his family will be able to return to Mexico and regain their citizenship there. For them, escaping the prejudices and the terror in the U. S., gives them a chance for a new life, one with dignity and security. For this, José and family will do the stone work, and quickly.

"Second, I have an arrangement with Joe Ablehammer, the man providing the barn raising crew, to set up a date between him and Lady Gaga as guests of honor at Houston Tuttle's inauguration."

"How does this not result in a lot of publicity?" asked Sloop.

"José's family are just one step away from government officials who want to arrest them for the freshly minted, treasonous act of forfeiting their U. S. citizenship. When they are successfully back in Mexico, the story of this barn will be of no interest.

"And Joe's crew are all immigrants as well. With the residues of the current economic and political climate, I doubt their stories will be pursued."

Shirley turned to Enno and nodded her head. Enno's nod was barely perceptible.

"How many people do you expect and how long will they be here?" asked Helga.

"Fifty or so for four to five days maximum. They'll be camping and I've got Jimmy helping me to round up food for them. Not sure about cooking and such, though."

Enno dropped his head, slowly.

"I think we'll have plenty of people to help on that front." Helga again.

Under a tree, Monk stood at the outer edge of the circle of friends knowing this silliness was grand, good and growing. Then he was surprised by the hand of his wife on his shoulder. Shirley caught the moment from across the way and blurted "Vicki?!"

Small men and short women of great learning and greater concern had been secretly holding international grand jury hearings in Copenhagen and were ready to file indictments in the world court charging RBMC and the Myrmecophilus Group with anti-trust violations, wholesale mental enslavement and gross violations of biological law. Very soon a long process would begin, the result of which would be the dismantling of global internet commerce and the exorcism of billionaires who show no actual working history.

Shirley and Monk hear of the Redding riot. Bob Roy Jeb trampled to death. Enoch Drub unmasked, crowd now going, en masse, in search of the real one. The royal one. *The Engine of Quanchee* released and biting itself in the butt.

mwmwmwmwmwmwmwmwmwmwm

On the ocean floor of Saturn's icy moon, Enceladus.
whose monkeyed futures lace the never to be told histories,
ice skeletons are energized by envy of brother Titan's methane atmosphere
sugar with no palate towards hunger no grab no flutter.

"Redding?" Margaret Curve is seen in the Target parking lot slamming the dash of her $172,000 BMW i8 Roadster with the flat of her hand. Slamming it over and over again. The people watching are imagining this middle-aged woman is nothing but a spoiled brat having a bad day.

Margaret logs in to her bluetooth phone, one of her realtor/enforcers on the other end. "I think Drub was a ruse. This is what I know: A local guy, his name is Enno Duden. He shows up in no data base. He's certifiably nobody. He's young and seems less than educated. He's camped at some remote property way out of town with an entourage and they are trying to help him build a farm, from scratch. He somehow has gotten control of over a thousand acres. I don't think he's going to be a problem. The land is useless and he's going to fail with his project. But he is gathering converts."

"He might be one of the ones we're looking for. Or maybe Drub is for real. Can't take a chance. We need them both gone. One way to stop them is money. This Duden has to have money to make his project fly. We prevent him from getting money and all of this just goes away."

On other end of call, "Our informant tells us he has no money, no real money. And he's done all of this so far without it."

"Are you telling me he has gotten land and is building a farm, buying equipment, all of this, and without money, no capital, no loans, no benefactors, no inheritance?" Margaret in disbelief.

The voice on the phone answers. "The farm is happening because Duden never gives up, never compromises, there was always just this goal. Because he shows it, you look at him, listen and look and you know he is going to make this happen."

"So people are just giving it to him because he demands it?" Margaret asks.

"He demands nothing. I don't know how to describe it. He's just there, on his mission, and people are following him and helping. More and more people every day."

Then this is the one, thinks Curve, the one we've been expecting and watching for. He has to be stopped. We have to end him.

She connects with Jay Bossweiner and tells him. His response?

"I think we have a better way. We are going to begin a social network blitz, talk up how this kid and his followers are blasphemy to established religions. How they are quietly claiming to have in their clutches the true Messiah. While in 'truth' he is just another false prophet. We start another holy war. With direction and luck, in no time we will have circles closing in on this Duden character. We keep him alive until he is fully discredited. Then they can have him."

Enno has been sentenced to death by social caprice. Society wants him dead because, on the one hand it sadistically thinks his death is required à la a witch trial to prove he is the chosen one. This requires a dramatic public execution, while on the other hand society wants him dead because what he represents messes with the need for status quo. This requires for the boy to just go away and never return again. To be assassinated, to be disposed of.

You and Ashwan are his attorneys. We all are witnesses, some for the defense some for the prosecution. And all the rest of you, the rest of humankind, are here and there to be entertained by the bloody drama of it all.

Prompto was doing his sweep search, teasing and forking apart the dark web, and picked up a top secret bulletin. Seems Homeland Security was excited to discover the name Enno Duden in a land sales data base for Santa Cruz, California. A farm was purchased in the name of a watch-list entry. Result: HS arresting Dr. Filipiano on a terrorism warrant which accused him of harboring a religious extremist/terrorist. They demanded he reveal Duden and his whereabouts. Dr. Fil thought about it for ten seconds and said no thank you. They slapped him for the first time. That sealed it as far as he was concerned. He had a passing thought of protecting the boy. but truth be known he was far more concerned that the 'guvmint' didn't get close to the Filenium deposit. They slapped him fourteen times more so he diverted his pain during torture. It was relatively easy, he just imagined each of the three men dressed in pinafores. They transported the good Doctor to their halfway house in Pico Rivera. He joined thirteen other 'detainees' in an abandoned motel ballroom where torture trainees watched instruction film, ate chicharrones and practised new nastiness. One of those trainees earned extra money slipping information through the dark web to P. M.

Prompto Manks then forwards the story to Sloop and sends an edited version to Margaret's inbox.

"I knew it! Snake Flats was a 'curve' ball. Everybody to Santa Cruz now!"

Bossweiner added the available twist to the social network tale: "Enoch Drub is dead, Duden killed him."

> *"But Captain Vere was now again motionless, stand-*
> *ing absorbed in thought. Again starting, he vehemently*
> *exclaimed, "Struck dead by an angel of God! Yet that angel*
> *must hang!"*
> *- Herman Melville, Billy Budd, Sailor*

In a narrative, if this be one, where the story ends too soon the author is stuck doing his or her own laundry. Perhaps not today, but eventually. For the reader, crumbs on the bed, disappointed and unable now to sleep, the only thing which will satisfy is the next low slow lie. Imagine what you will, that afterall is the only useful result of any good liar's long story.

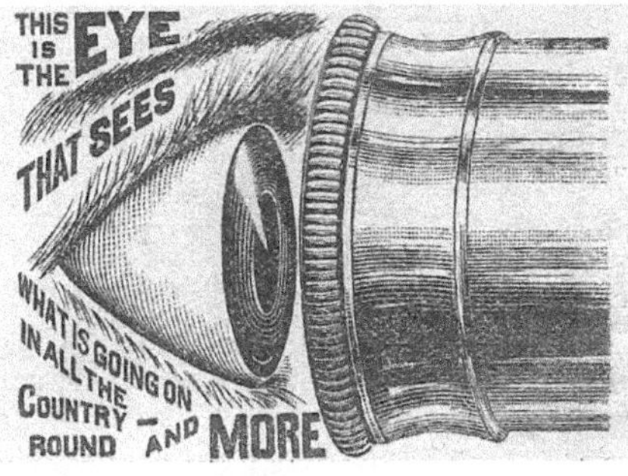

He just cannot accept that the story is over. The narrator has left his post, to insert himself head first into the romance of the fray, he has fallen victim to persuasions, though they are above his ability to introduce, yet and so he goes needing membership and consequence more than station.

The Dude will continue to abride.

So we flip him.

Now know him as Crude Eponymous Dirge. Though he prefers to hide behind the moniker Levi Tookus. Just don't worry about him. Think of him as an unfortunate coffee stain on the manuscript. Think of him as the mysterious memory of an indiscretion never fully realized though no less regretted.

<Yours is not the only ear within I might whisper. I will haunt you and you for a very long time.>

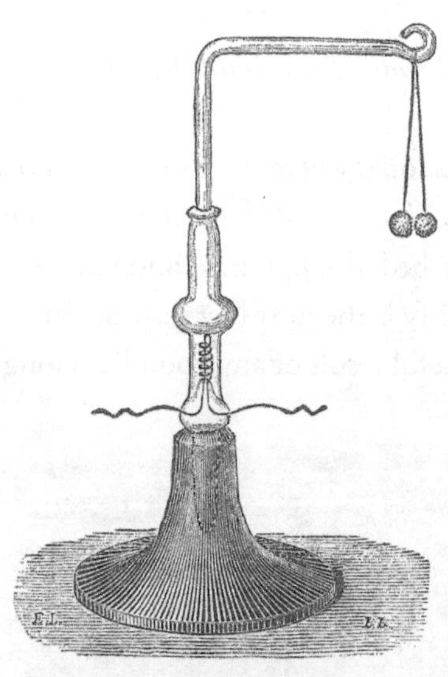

chapter twenty six

people gates

*"I write as if to save somebody's life. Probably my own.
Life is a kind of madness that death makes. Long live the dead
because we live in them."*
- Clarice Lispector, A Breath of Life

*The secret to life is to learn to love. Without this
capacity an argument could be made that humanity
is no longer sustainable.*

The eleven are present, as are the second, third, fourth and fifth eleven. They are raising a barn in fast frame, they are rock picking, they are inserting windows and doors in the honeymoon cabin, they are cooking and feeding, and bandaging, and joking, and teaching each other words in their languages. The Haitian carpenter and the Mexican stone mason are giggling. Benji and Ash are vigilant and thrilled. Jimmy and Jon are stirring massive cauldrons of Ecuadorian chili made with prunes, corn, pork sausage, white beans, tomatillos, pinto beans, small purple potatoes, molasses, black beans, achiote paste, tequila, anchovies and wild turkey (the meat) and wild turkey (the lube).

Shirley is everywhere, carrying both unbearable happiness and terrible dread. Every ten seconds she looks to anchor her soul with a glimpse of Enno, and most times beams when she sees his returning gaze. Pregnant Victoria and freshly frail Sig are at her side, offering themselves as soft crutches for this beloved woman who stumbles with her manic urgencies.

It is a rush and it is a rush.

Out along the edge the first wolves sniff forward towards the activities. The smell of the chili is driving them nuts. Resumé is going nuts too. First because that voice is still in his head, and second because the high pitched smell of the wolf anxiety is deep in his nostrils.

Well before his inauguration, Tuttle declares his intention to end Martial Law, to set up a constitutional convention to reinvent and rewrite his country's governance, to demand a testing and licensing process for journalists, to implement a re-arming of antitrust laws, to fully outlaw chemical farming, to provide free health care for everyone and food for everyone, to build cooperative canning facilities in every rural town, to establish an entire new education system with state funded schools... While the present day journalists complain vehemently and incessantly about the failed bargain of this newly elected president. How dare he wait until after he was elected to finally come clean on his agenda? And Tuttle frequently responds, "Phooey on you."

In the Brown Dwarf cavern, the ghost of Titus is haggard and woozy, like a sixteen year old girl in the final fifteen minutes of her first night of active flirtation and dancing. He is still spinning and floating but every once in a while, when his axis drifts out to the edges, he slowly descends to touch the floor, and when he does flashes of color return to his black, white and grey form as electricity fills him. And with this he feels a long lost sensation, he feels hungry. So he does the only thing he knows how, it's his protest and his cry for help, he thinks as loud as he can the word "Dog."

The last rafters are going up and the clouds are racing by. Enno has found himself drawn to the ridge pole, helping to align boards and hammer. Then he is surprised by Sloop who has climbed the roof to hand a small pine tree branch to him just as the last rafter is nailed. "Let's plant this on the ridge together son? The bell is ringing and they have supper ready for us all." Enno was in tears, his right hand on Sloop's shoulder and Sloop's on his. The clouds parted and Sloop descended. "Be right down." Enno stood on the ridge listening, feeling. Something was happening. Sunlight poured through the clouds backlighting Duden as the entire crowd of workers and friends looked up to see his form. He stood upright head back, listening, arms out slightly from his body as though balancing.

Everyone was standing looking up, everyone shivering because of the revelation, only the breeze through the pines made any noise. Then Resumé started barking and a flare from Nimbo's signal gun cut across the sky in colored smoke. Shirley grabbed her mouth and started to run, she had no clue where she was going only that she had to get there. She stopped when she saw Enno throw then scatter himself down, precariously, down the roof framing, heading down to the ground, he wasn't falling, he was transferring each slip to a positive grab and thrust, doing some sort of monkey act.

The first pack of wolves had been joined by others who were backed by wolf men and crazed forms, carrying sticks and pitchforks. Mikey, deGaulle and Sarkozi had got Nimbo's signal and went wide to hold the wolves at bay. Sloop signaled Monk and was joined by others as they armed themselves in preparation for an assault. Enno had disappeared. When he hit the ground he was met by Benji and Ash who were led by a nervous barking Resumé. No words were said. The little men grabbed whatever hand tools they could and followed Duden and his dog. Enno was muttering to himself, 'why haven't I been listening to you?' And Resumé heard him now finally listening. They went as living rockets towards the cave.

Shirley shook off all her 'worry' nonsense and went into clarity.

Enno had felt the tremors when he was on the ridge pole. Now everyone else could feel them. The ground was preparing itself for an earthquake or something disastrous. All Enno knew was that he had to get to the cavern right away. Following, Shirley had arrived at the box canyon too late, from a distance she watched Enno, Resumé, Benji and Ash enter a crack in the rock wall only to be followed moments later by a large but soft explosion that filled the air with a nasty methane smell.

It was Jimmy who observed "Wow, the earth just farted."

There were three crowds following Shirley. First there were the five elevens. And those were followed by a mob of crazed strangers wielding sticks and axes. They were backed up by four large packs of wolves wearing radio collars. The wolves stopped and seemed a little disoriented, sniffing all around. Then the lead wolf turned and ran back to the barn site. He had finally locked in on the smell of that unattended, cooking chili. All his fellow beasts followed him.

The explosion at the cave entrance had been followed by a wind, then dust. Someone hollered "cave in!" Sloop, Lloyd, Jackson, Jon, Jimmy, Shirley, Sig, Smudge, and Helga ran into the cave prepared to dig them out. There were three more explosions, more dust, more stink, more wind. The crowd waited, none among them wanted to chance following. Most of them sat in a big ring on the ground, complaining, crying, praying and wondering.

Back at the barn the wolves had consumed the chili that had been made to feed 75 people. Their eyes went even redder and they stumbled off to confuse the wildkife biologists who were going to have to interpret the radio signals emitted from collars on stumbling bloated wolves.

The earth went quiet, the clouds moved over and there was a sense of doom. A couple of helpers got up to go to town and report the catastrophe when someone said, "What's that?" Stumbling out from the crack in the rock came each and every one of them, Resumé first. He seemed to be grinning. As they came out they just naturally formed a choreographed half circle, each looking back at the entrance to see who would come out next.

Shirley moved ahead to receive Enno, as he came out cradling a tall, thin, grey-haired old man.

It was Jimmy Three Trees who announced with no stutter, "It's a miracle! Titus is alive!"

Enno laid the old man down near a big stone and Titus sat upright. Sloop and Helga went to him. Every one of the eleven went down on one knee to be at eye level with Titus and to congratulate and interrogate him. Everyone except Enno who stood and faced the crowd. Every single one of those watching went to a knee even those of the mob who, before they did, relaxed and set down their weapons.

The clouds raced off, sunlight exploded on him and Enno stood. Shirley joined him.

Within these end times, the less than beautiful more than young, those alone enough with their own breathing, able to hold the pain of others, seldom familiar with strangers, ever respectful of the wait, they these know themselves as forest. The snow melts, the winds hurt deep, the shedding hairs hasten all watching for a better next, for a splendid difficult formula.

These others know end days to recycle - they fear not the end but instead the moronics of empty repeat. It is the seasons they crave, seasons purposed.

And so it is come
that one arrives and answers
- answers completely.
Or is that simply what we want to believe?

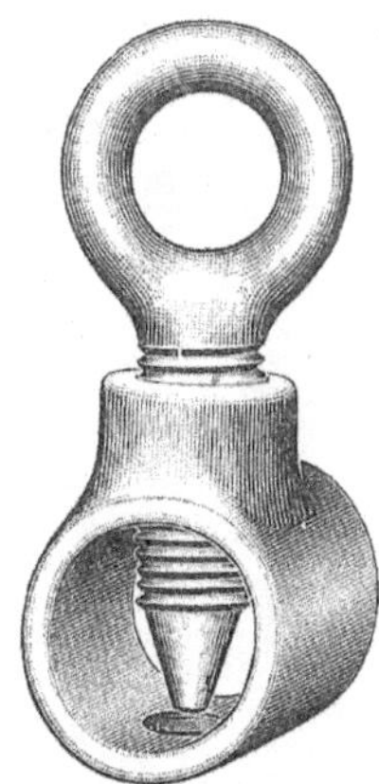

"I am going to seek a great perhaps; draw a curtain, the farce is played out."

\- *Rabelais*

Brown Dwarf Character List

Enno Albert Duden
Gendun Drub aka Enoch Drub
Resume´
Benjamin Aboo aka Benji
Titus Ibid
Ashwan Clevalure aka Ash
Shirley Intoit
L. J. (Lloyd Jerald) Shoulders
John B. "Sloop" Ogdensburg
Jimmy Three Trees
Sigismundo Maltesta aka Sig aka The Monk
Uncle Nimbo aka Jim Nimbus Sparkle
Jon Drummer
Dude Ironymous Binge aka Crude Eponymous Dirge
aka Levi Tookus
Helga Ogdensburg
Jackson Fork
Dr. Fil (Dr. Filipiano Catalgo Magneto Internuncio)
Vicky Wood aka Victoria Maltesta
Anatoly Grim
Vic Nib
Olivia Perchance
Margaret Curve
Cash Apercu
Elfin Flit (dog Penny)
Turk Swoop
Smudge Cochran
Leonid Runaroundovitch aka Leo Vitch
Bessie Quince
Loraine Frill
Penny Frill
Gunny Sax   aka   Gunnison Tennner Sax
Houston Tuttle
Filson Wool
Benny and Swirl Hubris
Bob Roy Jeb
Jim Parks aka Promto Manks aka PM

Doobie Lexicon
Jimmy Sturgeonburger
Besom Cellar
The Engine of Quanchee
Plumley & Recto
Noah and Sylvan Vibbert
Stu Jeff
Bart Prussian
Hardlee Nuff
Dave Actualroger
Ulrika Freya Orley
Tronk
Dr, LaLouche
Wilkens Boys, Jimmy and Bobo
Hector and Ilbana
Jose Jimenez
Joe Ablehammer
Jay Bossweiner
Peter Schwartzwarden

Guest Appearance: Jeff Goldblum

What holds all this together?
Devotion.
What keeps it from flying apart
from losing track, from silliness?
Balance.

The old man's brain was a small, forgotten, dusty train station,
the ideas went flying by, sometimes with a horn blast.

*Far back in the last century, the author
riding the board, Gary Eagle driving.*

Antikythera Mechanism

roots in a lovely filth

book three - duden chronicles

Lynn R.Miller

2022